Environment and Society

Environment and Society
Advances in Research

Environment and Society

Advances in Research

Volume 3 - 2012

■ EDITORS ■

Paige West and Dan Brockington

berghahn
journals
NEW YORK · OXFORD

Environment and Society: Advances in Research presents articles and critical reviews reflecting the work of environmental anthropologists, geographers, scientists, human ecologists, and scholars in other environmentally oriented social sciences from all parts of the world. The field of research on environment and society is growing rapidly and becoming of ever-greater importance, not only in academia but also in policy circles and for the public at large. Climate change, the water crisis, deforestation, biodiversity loss, the looming energy crisis, nascent resource wars, environmental refugees, and environmental justice are just some of the many compelling challenges facing society today and in the future. Through this journal, we hope to stimulate advanced research and action on these and other critical issues and to encourage international communication and exchange among all relevant disciplines.

COPYRIGHT

ONLINE

Environment and Society: Advances in Research is available online at www.journals.berghahnbooks.com/air-es, where you can browse tables of contents and abstracts, purchase individual articles, or recommend the journal to your library. Also visit the *Environment and Society* Web site for details including full contact information for the editors and complete submission instructions for contributors.

ADVERTISING

All inquiries concerning advertisements should be addressed to the Berghahn Journals editorial office: advertising@journals.berghahnbooks.com.

Environment and Society
Advances in Research

————■————

Contents
Volume 3 ■ 2012

Introduction
Capitalism and the Environment

Paige West and Dan Brockington

Capitalism is the dominant global form of political economy. From business-as-usual resource extraction in the Global South to the full-scale takeover of the United Nations 2012 conference on Sustainable Development in Rio, Brazil by corporations advocating the so-called green economy, capitalism is also one of the two dominant modes of thinking about, experiencing, and apprehending the natural world. The other dominant mode is environmentalism. There are many varieties of environmentalism, but the dominant mode we refer to is "mainstream environmentalism." It is represented by powerful nongovernmental organizations and is characterized by its closeness to power, and its comfort with that position. This form of environmentalism is a well-meaning, bolstered by science, view of the world that sees the past as a glorious unbroken landscape of biological diversity. It continuously works to separate people and nature, at the same time as its rhetoric and intent is to unite them. It achieves that separation physically, through protected areas; conceptually, by seeking to value nature and by converting it to decidedly concepts such as money; and ideologically, through massive media campaigns that focus on blaming individuals for global environmental destruction.

Contemporary capitalism and contemporary environmentalism came of age at the same time. The extensive global decolonization movements in the 1960s and early 1970s altered the ease by which capitalists and corporations could access new sites for natural resources, land, and labor; the three key ingredients for keeping capitalism growing. This, coupled with the oil crisis, and the realization that access to cheap and easy oil—the commodity that drives capitalist expansion—could no longer be taken for granted, ushered in the age of flexible, highly mobile capital that we have today. The next decade gave rise to corporations that were lean and seeking deregulated environments from which to draw resources. If they could not have open and free access to natural resources, land, and labor through collusion with colonial oppressors, they would seek to influence new, and old, nation-states, to deregulate access to everything.

The global environmental movement, while having roots in the nineteenth and early twentieth century preservationist writings of Henry David Thoreau and John Muir and conservationist writings of Gifford Pinchot, also came to maturity in the 1960s and early 1970s. *Silent Spring* was published in 1962, *The Limits to Growth* was published in 1972, and that same year the crew of the Apollo 17 spaceship took the first clear picture of an illuminated earth from space. Also in 1972, the United Nations held its first conference on the environment, bringing together governments from both the so-called "developed" world and the newly decolonized states. These events ushered in the decade when the United States and other global powers passed environmental legislation at an unprecedented scale (e.g., the clean water act, the endangered species act, and the clean air act in the United States).

Environment and Society: Advances in Research 3 (2012): 1–3 © Berghahn Books
doi:10.3167/ares.2012.030101

The early environmental movement was one that directly challenged its age mate—capitalism—with critiques of the corporate and state-driven disasters from Bikini Atoll to Three Mile Island, as well as with serious public education campaigns around pesticide use, acid rain, environmental racism, and the overconsumption of oil, gas, and electricity in the Global North. But in the 1980s, environmentalism took a step back from a posture of radical critiques of corporations, states, capitalism, and the collusion between the three, and began to focus its energy on the poor people living in highly biologically diverse places that seemed to be on the edge of capitalism.

Environmentalism went south, so to speak, and inserted itself into the power struggles over environmental governance in the recently decolonized nations. While there, it got snugly in bed with its old enemy, corporate capitalism. Because environmental organizations wanted to usurp the power of state regulatory agencies, based on their assumption that they knew better how to conserve biological diversity than did peoples in the Third World, and because corporations wanted deregulation so that they could continue to find cheaper and cheaper access to land, labor, and natural resources, they made perfect partners.

We live in a world where almost nobody with real international power challenges corporations, their actions, or the logic of contemporary capitalism. Mainstream environmentalists have in some cases, for example in places where gold mining or hydroelectric dam construction have destroyed important ecosystems, taken a back seat to corporate power in exchange for state promises to conserve other so-called "valuable" areas. In other cases, they have brought corporate leaders directly onto the boards of directors of their organizations and have taken gigantic grants from corporations, always arguing that the money in no way makes them less effective or unethically accountable to corporations. It is rare today to see a large-scale and powerful environmental organization challenge corporations or their logics. Indeed, the environment has become just another vehicle for capitalist accumulation and, mostly, it feels that there is nobody there to stop this.

The six papers in this edition of *Environment and Society: Advances in Research,* attempt to clearly describe the contemporary relationship between capitalism and the environment by reviewing five distinct and important literatures in the social sciences.

In "Dollars Making Sense: Understanding Nature in Capitalism," James G. Carrier steps away, slightly, from the standard review-style article we publish in *Environment and Society* in order to analyze the connection between capitalist enterprises and people's understandings of things that happen in and happen to what they perceive as "the environment." Carrier works through how commercial pressures make certain activities, specifically those intended to alleviate hunger and those intended to conserve the environment, present problems and humans' surroundings in particular kinds of ways. He then shows how these representations are simplified in ways that are meant to encourage certain forms of thought and action.

Two of the articles attend to questions about neoliberalization and the environment. In "Neoliberalism and the Production of Environmental Knowledge," Rebecca Lave works through the literature on environmental appropriation, commercialization, and privatization in order to connect these processes to environmental science. Drawing on a wide interdisciplinary range of work, she carefully shows how neoliberalism—as a philosophy and as a set of policies—affects knowledge production both inside and outside the academy. In "Fisheries Privatization: Capitalist Logics and the Remaking of Fishery Systems," Courtney Carothers and Catherine Chambers connect multiple processes of commoditization with the social and material production of the human-marine relationship. Specifically they work through the literature on privatization to show how the marine world comes to be envisioned and acted upon by both businesses and conservation-related actors.

The next two articles address tourism as a vision and practices that bring together the ecological and economic realms for both capitalists and activists. In "Contradictions in Tourism: The Promises and Pitfalls of Ecotourism as a Manifold Capitalist Fix," Robert Fletcher and Katja Neves examine the relationship between ecotourism ventures and capitalist ideology, practice, and structure. In particular, they explore the role of ecotourism in attempts to solve some of the central contradictions in capitalist accumulation. Drawing on a wide range of interdisciplinary literature they argue that ecotourism, when advocated as a development fix for poor people living in highly biologically diverse places, can be seen as an endorsement of particular forms of "fix" to these inherent problems and contradictions. In "From a Blind Spot to a Nexus: Building on Existing Trends in Knowledge Production to Study the Co-presence of Ecotourism and Extraction," Veronica Davidov shows how ecotourism and multiple forms of research extraction not only exist in the same places at the same times but also how they coexist in what she calls a "nexus." With this, she asks why this coexistence, and in some cases a reliance on the same social and ecological histories, has been underanalyzed in the social sciences and how we might work to understand this phenomena both methodologically and theoretically.

Finally, Yda Schreuder's article, "Unintended Consequences: Climate Change Policy in a Globalizing World," connects capital, climate, and policy. Schreuder examines the effects of the cap-and-trade system in the European Union as one example of the unintended consequences of capitalist responses to environmental problems.

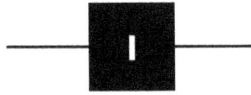

ARTICLES

Dollars Making Sense
Understanding Nature in Capitalism

James G. Carrier

■ **ABSTRACT:** This article addresses the relationship between enterprises in capitalist systems and people's understandings of activities concerning the environment. Two sorts activities are described, those intended to alleviate hunger and those intended to protect the environment. Both illustrate how the routine operation of those enterprises affects the ways that people perceive the world and problems in it, and how people are likely to evaluate activities that can address those problems. Such effects come about because normal commercial pressures make it likely that enterprises will present the surroundings and the problems that concern them in ways that stress certain aspects of and processes in the world, while slighting others. The result of those presentations is a simplified rendering of the surroundings that tends to encourage certain sorts of orientation and action rather than others. The relationship between these renderings and those orientations and actions is not, however, straightforward, and this article concludes with a consideration of the sorts of processes that can shape that relationship.

■ **KEYWORDS:** capitalism, environment, ethical consumption, Jamaica, legibility

The economic crisis that began in 2008 has led to a renewed interest, in both popular and scholarly media, in economic practices and institutions (for scholarly work see, e.g., Ho 2009; Ouroussoff 2010). In the popular media, much of this interest has centered on activities and individuals who can be cast as deviant or fraudulent, such as "rogue traders," bank employees who make unauthorized trades, and companies that misrepresent the financial instruments and services that they sell. This focus on the deviant in the commercial world is not new (Eichenwald 1995) and can be revealing. However, it tends to divert attention from the norm, the routine pressures on and practices of conventional enterprise in capitalist systems.

The purpose of this article is to approach the title of this special issue, "Capitalism and the Environment," in terms of those pressures and practices. In some ways, such an approach is familiar: we all know about the drive to profit and the ways that it can affect the surroundings, people's relationships with those surroundings, and people's lives. The commercial pressures that operate in the petroleum industry, for instance, can lead to changes in the procedures used for drilling and the places where that drilling occurs. If these changes are successful, they

Environment and Society: Advances in Research 3 (2012): 5–18 © Berghahn Books
doi:10.3167/ares.2012.030102

become the innovations that competitive capitalism is said to encourage. Alternatively, they can fail spectacularly and result in things like the disaster of the Deepwater Horizon in 2010, the Bhopal Union Carbide disaster in 1984, and the wreck of the *Torrey Canyon* in 1967.

Pursuing those ordinary pressures and practices is worthwhile. For one thing, focusing on them balances the tendency for people's attention to be drawn by the spectacular, whether the rogue trader or the lethal drilling rig, and hence drawn away from the nature of the system in which these events occur. As well, such a focus makes it easier to ask certain questions. If we start with the nature and operation of the system of constraints and pressures in which enterprises operate, we can work out how these are likely to be manifest. This allows a perspective on the relationship between capitalism and the environment that is different from what we would have if we started with the failures and worked inward.

The mundane pressures and practices of concern here can be approached in different ways. Borrowing from Merton's (1968) idea of theories of the middle range, the approach here is a focus on processes of the middle range. That focus is on a level below the macroscopic, the realm of the broad structures of political economy and historical change that have been used to produce revealing models of contemporary renderings of people's understandings of and relationships with the natural surroundings (e.g., Argyrou 2005; Escobar 1999; Ingold 1988). Equally, however, it is above the level of the microscopic, the realm of the detailed processes and procedures that are the concern of fields like the social studies of finance (e.g., Lépinay 2011; Muniesa 2007; Zaloom 2006). Certainly the middle range is shaped by the macroscopic and the microscopic, but attending to it can help illuminate the ways that the pressures and practices that operate at that level can have a significant effect on people's relationships with their surroundings.

I pursue that topic in different ways in the three main sections of this article. In the first section I present the basic analytical approach that I use to discern the ways that dollars make sense, the ways that routine commercial concerns and practices shape the ways that people understand the natural world, in terms of the use of genetically modified organisms (GMOs) as seed for staple grains. That presentation is normative and somewhat idealized, in Weber's sense, which makes it easier to discern the tendencies that inhere in the routine orientations of the sorts of people and companies I describe. The second section shifts from the idealized to the concrete, and takes cognizance of the contingencies that were ignored in the first section. It does so with a discussion of the ways that two national parks in Jamaica present the areas of coastal waters that they seek to protect. The third section qualifies the concern of the first two, which is the constructions of the world that people confront as a result of routine commercial constraints and practices. That qualification is necessary if we are to avoid two errors. One is the error of assuming that people accept and adopt those constructions in a straightforward way; the other is the error of assuming that when enterprises produce representations they do so without reference to the ways that they are likely to be perceived. To show how these assumptions are faulty, this third section points out the broader social and cultural factors that can affect the way that people interpret those constructions, and the routine commercial pressure on enterprises to take those factors into account.

The Power of the Purse

I turn now to my first task, laying out in fairly idealized terms the analytical apparatus that allows us to see the ways that dollars make sense. I do so with an illustration that revolves around the power of the purse. That power is the appropriate place to start, for money is the crucial resource for enterprises of all sorts in capitalist systems with extensive market economies. That illustration concerns genetically modified (GM) varieties of staple grains.

We are told regularly that the amount of food in the world is decreasingly adequate for human needs, perhaps because of rising population, perhaps because of climate change, perhaps because of changes in people's diet as they become wealthier. Whatever the cause, the result is dire: a global rise in the cost of food and a growing threat of hunger and civil disorder.

This vision of looming hunger often is invoked by those who advocate GMOs, and they argue that GM crops will allow an increase in our food supply (such arguments are considered in Stone and Glover 2011). In the words of a commentary in the *Telegraph* (Emmott 2008), "The longer we deny ourselves this technological way to increase food output …, the longer the current imbalance between food supply and demand will last." The advocates' argument can be challenged (famously by Sen 1981; for an overview, see Stone 2010), but my purpose here is to trace its implications, rather than assess its accuracy. In particular, I am concerned with the type of understanding, the sort of sense, that the argument of those advocates solidifies, and the type that it slights. I describe these types by means of another argument, also invoking straightforward aspects of food and of commercial life.

The problem that the advocates of GMOs point to is one of insufficient food. However, the problem of insufficient food is not as simple as it might appear at first glance. It can be taken to be the problem that the advocates identify, that is varieties of grain that are insufficiently fruitful. Equally, however, other things contribute to that fruitfulness. These include soil composition, water supply to the crops in the fields, and agricultural technique. As well, there is crop storage and transportation, and all the other steps between the point where the grain leaves the field and the point where it becomes the focus of concern, food that is in people's stomachs.

These different ways of rendering the problem of insufficient food point in different directions, and so encourage different understandings of the food supply and its uncertainties. For instance, it appears that a substantial amount of food is lost in storage and transportation between harvest and the market. The Food and Agriculture Organization (1989: n.p.) says: "Estimates of the post-harvest losses of food grains in the developing world from mishandling, spoilage and pest infestation are put at 25 percent," while the African Post Harvest Losses Information System (n.d.) puts the figure for grain loss in sub-Saharan Africa at between 14 percent and 17 percent.

If these estimates are correct and these losses were reduced by about half, we would increase the food that could go into people's stomachs by about one-tenth, which is substantial. If a significant proportion of that loss is from vermin and spoilage, then measures such as ventilating grain stores to reduce spoilage and modifying them to keep vermin out would go some way toward addressing the problem of insufficient food. Further, it is likely that similar changes between planting and harvest also would help. In drier areas, for instance, it might be possible to improve irrigation systems by things like increasing the lining of irrigation ditches, thereby reducing the amount of water lost through seepage. Other such improvements will be apparent to those who are familiar with agriculture in different parts of the world (see Hillel 1997).

I have presented some of the ways that the problem of insufficient food could be addressed, ways that strengthen certain understandings of food and slight others. In focusing on one element in the long and complex process of food supply, they encourage an abstract and simplified understanding of the goal that they set themselves, getting food into people's stomachs. The argument for GM varieties, like the argument for IR8 rice in the Green Revolution and the argument for any other improved seed variety, stresses the initial link in the chain from seed to stomach, and in doing so it slights the others. Likewise, the argument for making grain stores more resistant to vermin stresses a different link and, again, slights others.

There are, then, a variety of perspectives on the problem of insufficient food, each of which renders that problem in its own terms. Given this variety, it is pertinent to ask why some appear more visible more often than others. In terms of my illustration, why is the advocacy of the expansion of GM crops so much more common than the advocacy of, say, increasing the production of clay to line irrigation ditches, of sheet metal to line grain stores, or of wire mesh to allow better ventilation? This question is important because the more visible advocacy is more likely than the less visible to influence general public understandings of things like "insufficient food" and how to correct it. In saying this, I treat the sheer repetition and visibility of the case for GM crops as an aspect of what Steven Lukes (1974) calls the third form of power, the power to define what the question is and what answers to it look like. It is the power to encourage people to make sense of things in certain ways rather than others, and hence to induce people to act in certain ways rather than others, including acts that are more favorable to the spread of GM crops than they would be otherwise.

Understanding why the advocacy of GM crops is so visible means attending to the power of the purse. That power can be put simply: large commercial interests benefit from the increased use of GMOs, in a way that they do not with an increase in clay, sheet metal, and wire mesh. Because many people associate "large" with "bad," I must stress that identifying these commercial interests as large does not mean that they are involved in deceit. They may be deceitful, but as I said, here I am concerned with a normative state of affairs, not one characterized by deceit or deviance. Companies like Monsanto invest substantial resources in developing new products, which include new varieties of grain that, when used in the appropriate ways, will produce significant increases in the amount of food harvested. However, these are large companies with substantial resources. The consequences help illustrate the power of the purse.

One of the consequences springs from the context in which those new varieties exist, a crucial aspect of which is patents. When a company like Monsanto develops a new variety of, say, rye, it will patent this variety, and the patent allows it to extract rent from it. This rent can be seen as a return on the resources spent developing that new rye, but the patent frees the company of competition from alternative suppliers and allows it to charge more than it could otherwise. Developing a new variety of rye is expensive, and rewarding and encouraging innovation is the reason given in the US Constitution (Art. 1, §8) for patent and copyright. More is involved, however, than rewarding the diligent inventor. For a large company like Monsanto, that reward is important because it allows the company to pay what it owes to its bankers and bond holders, pay dividends to its shareholders, and maintain a high price for its shares. As this indicates, the large commercial interests at issue extend beyond Monsanto itself, and at the more macroscopic level they extend, ultimately, to the reproduction of the capitalist institutions and relations of which Monsanto is a part. To secure that reward, we can expect that Monsanto will, naturally, identify and tout the valuable things that their rye can do, including combating world hunger.

The world of clay, sheet metal, and wire mesh is different. These technologies lack the allure of innovation; no one writes stories about them for newspapers. Rather, they are what those in business call "commodities," fungible bunches of utility or use value: a square meter of galvanized iron 1.5-millimeter thick is pretty much the same as any other square meter of such galvanized iron. Companies do not want to have their products be commodities, for it reduces their potential profits by driving them to compete with other producers on price and quality. Of course companies that produce galvanized iron do so for a profit. However, such companies are in the relatively weak position of those who produce commodities. They cannot secure the formal rents that come with a patent, and because their products are fungible they find it difficult to encourage demand for their particular product through things like advertising, which would allow them to secure the informal rents that come with a popular brand name. The result is that there are few powerful commercial interests around to advocate sheet metal as a solution to world hunger, and hence little encouragement for the public to make sense of things in the appropriate ways.

The purpose of this section was to lay out my basic approach, and I have done so by pointing to the sorts of processes that deserve attention if we are going to understand the role of dollars in making sense, the ways that the routine operation of companies in capitalist systems can influence people's understandings of their surroundings in general, and the environment in particular. In this case, the routine commercial interests of companies like Monsanto make it likely that they will seek to generate demand for their seed varieties, and the context of the United States Patent and Trademark Office makes it likely that those varieties will be patented. Because companies like Monsanto are powerful, and because the institutions that hold their stock and debt are likely to be powerful, they have access to the power of a very large purse to make the case for the virtues of those seeds. All that Monsanto and those holding their stock and debt need to seek is visibility, awareness among the public of the virtues of the GM rye. As I said, this visibility entails stressing a certain aspect of the problem of inadequate food, and a concomitant slighting of other aspects of the problem, and of other ways of dealing with it.

However, and to anticipate a point I consider at greater length later, the power of the purse to generate visibility is only the ability to send a message. It cannot dictate the way that people interpret that message. Other factors that are beyond the immediate control of companies like Monsanto are important, though here too they have an advantage over those producing galvanized iron, wire mesh, and clay. As James Scott (1998) describes, one aspect of Modernity is the tendency to see things independently of their context. In this, Scott points to the value placed on abstraction in Modern societies (Carrier 2001), and advocating GM grain as alleviating hunger abstracts the genetic constitution of the seed from the context by means of which it becomes food in people's stomachs. As well, the advocates of GM grains benefit from the fascination with the new, with innovation, that is important in many Western societies (e.g., Edgerton 2008), and perhaps even from the common assumption, also a part of Scott's Modernity, that substantial problems, which include world hunger, are best addressed by coordinated, substantial programs of the sort that, it seems, only large organizations can carry out, rather than by the piecemeal local, conventional efforts that would lead to a piece of sheet metal here, a square of wire mesh somewhere else, and so on.

The routine pressures on and practices of enterprises in capitalist systems, my concern in this section, affect the most conscientious and honorable. A company may develop a new variety of grain expressly in order to combat hunger; that variety may in fact be markedly more productive than the other varieties available; the resulting publicity may be justified. Even if all this is granted, three things remain important. First, the company that developed the grain needs the profit arising from sales if it is to survive and prosper. They are unlikely to spend their money to develop and market varieties of grain that they think will lose them money, or that will profit them materially less than the other possible ways that they could use their money. The second is that the more resources the company has, the more it is able to generate publicity for its products, and so secure those sales. The third is that its publicity encourages one definition of the problem rather than others. Dollars, then, affect how people make sense of their world (descriptions of some of the complexities of this process in pharmaceuticals are in Applbaum 2006, 2009).

Ethicality

In the preceding section I provided a fairly idealized sketch of the ways that dollars make sense, the ways that the routine operation of enterprises in capitalist systems can affect the ways that people understand aspects of their surroundings. In this section I shift from an idealized to a concrete case, one concerned with much smaller issues and interests: two national parks in Jamaica and the bodies of water that they oversee. Moreover, my focus is not the words that people use to define particular aspects of the world, but the images that these parks use to portray the coastal waters. Those parks are broke, and they use the images as a way to generate resources.

Those parks are engaged in ethical consumption commerce, which develops as part of the emergence of a market demand by ethical consumers (Carrier 2012). It is the commercial activity intended to attract those consumers, who are concerned especially with the ethical aspects of what they purchase. Although much of that activity is carried out by companies, many charities and other organizations seek to appeal to ethical consumers of different sorts, and what motivates them is the same as what motivates companies: they need the money if they are to survive, much less prosper and achieve their goals. The two parks that concern me are in that situation.

Ethical consumption commerce is problematic, because it contains a contradiction. One aspect of this is that the organizations involved in it seek to appeal to people who are likely to be opposed to what they see as the heedless irresponsibility of many companies toward the natural (and social) environment, and to what they see as the calculative, impersonal rationality of the economic realm more generally, which includes the routine pressures and processes of commercial activity. The other aspect of this contradiction is that this appeal to ethical consumers relies on and expresses the logic of the economic realm that those ethical consumers distrust. In other words, those ethical consumers are being asked to express their disapproval of conventional economic practice by means of conventional economic institutions and relationships (see, e.g., Moberg and Lyon 2010). If the growing sales of ethical products is any indication, those ethical consumers are happy to do so.

This case is concerned with a particular consequence of that contradictory commercial activity, what I call "ethicality" (Carrier 2010, 2012), which I draw from aspects of the work of Scott (1998), especially as they have been elaborated by Errington and Gewertz (2001). Recall that one of the things that concerned Scott was what he called "legibility," making one or another abstract entity visible and recognizable, whether it was the revenue potential of trees that was made legible by the practices of Prussian state foresters or the modernity of Brazil that was made legible by the construction of the new capital, Brasilia. Errington and Gewertz (2001) drew on Scott when they described how the courts in Papua New Guinea identified specific social practices as instances of the legal concept "traditional culture." In other words, Errington and Gewertz were concerned with how the abstract entity "traditional culture" was made legible.

Ethicality resembles legibility, but the abstract entities that are made visible and recognizable are the ethical values that concern ethical consumers, such as "environmentally sound," "not exploitative," and "healthy environment." To a degree this resembles what I described in the previous section. There, the abstract entity was something like "world food problem," and it was made legible in terms of the fruitfulness of different varieties of seed. And just as the routine interests and activities of companies like Monsanto influenced the legibility of the world food problem, so the routine interests of firms involved in ethical consumption commerce influence ethicality, the legibility of those ethical values.

The Jamaican parks that I use to illustrate ethicality are in Negril and Montego Bay, which I studied intermittently from 1997 to 2005 (Carrier 2010). These parks received relatively little money from the Jamaican government, and sought to generate revenue through "user fees," typically fees charged to businesses in the tourism sector that wanted to operate in park waters. Thus, even though they were national parks, they were in the same economic position as conventional companies, for they needed to attract customers. One important way that they sought to do this was through their websites (for Montego Bay: www.mbmp.org; for Negril: http://negril.com/ncrps/), which presented information about the parks' histories and localities and the waters within their boundaries, and here I am concerned with the websites as they existed until about 2008, when these parks were some twenty years old.

Those sites included a number of images, typically underwater photographs of marine life. The accompanying text may have invoked concepts like "environmentally sound" and "healthy environment," but it was the images that represented those abstractions in concrete form, and so made them visible. Those who were knowledgeable about Jamaica's coastal waters could assess the validity of the way that these images represented those abstractions. However, as I describe below, the two parks were in resort towns that survived on mass tourism, so that the vast majority of the people for whom these images were intended were not likely to be very knowledgeable.

For such viewers, these images not only made those abstract concepts visible, they made them recognizable and so, ultimately, defined them. They did so because they were images presented by what appeared to be fairly authoritative and disinterested bodies concerned with the protection of Jamaica's coastal environment. These were, after all, national parks with the oversight of areas of coastal waters, not amusement centers. The difficulty is that these parks could not afford to be disinterested, however much their staff would have wanted to be. Rather, as I said, they resembled conventional amusement centers in Jamaica because they needed the business, and the images that they used were thought likely to appeal to potential visitors. In a word, they were advertising. This has consequences for understanding the ways that dollars, or at least the need for dollars, can make sense. Two examples will illustrate this.

The first example is fairly straightforward, and concerns what sorts of living things are portrayed in these images and what sorts are not. Overwhelmingly, those images are of things that someone diving in the waters would want to see: attractive fish and coral growths. Healthy fish and coral are a sign of healthy coastal waters, but hardly the only ones, and some of the others are less attractive to potential visitors. Prime among these are beds of sea grasses, which are important for nutrient cycles in coastal waters and as food for some of the creatures that live there. Although they are important aspects of healthy coastal waters, they appear to repel significant numbers of people, which accounts for why the company building a beachfront hotel in Negril cleared the sea grasses from the shallow waters that were intended as a swimming area for guests.

It is important to remember that these attractions and repulsions are the result of a range of factors that have shaped the image of the hedonistic tourism and its surroundings that Jamaica's resorts sell. These range from the nature of work and life in the United States, which is the origin of the vast majority of tourists in Negril and Montego Bay, to the renderings of nature, its pleasure and dangers that circulate in that country, renderings that include these parks' web sites and the advertising of Jamaica's all-inclusive hotels. Although such factors affect the tastes of potential tourists, as do the sorts of factors that I describe in the third section of this article, they are of little interest to those in charge of these parks and hotels. Those people are concerned with what those tastes are now, and as several people in the tourism sector put it, "People come to get away. They just want fun in the sun and the water … [they] don't want to know in depth" about Jamaica.

As I said, this example of fish, coral, and sea grass is fairly straightforward, as is what it indicates about the way that the concern to generate business can lead organizations to make a sound environment legible and, through that legibility, define it. In this case, commercial pressures lead to the use of images that include the attractive and exclude the unattractive, even though a sound coastal environment calls for both. The result resembles what I described of GM grain: a complex abstraction, whether world hunger or healthy coastal waters, ends up being simplified when it is rendered in concrete form, whether improved seed constitution or striking fish and coral. In both cases, making these abstractions legible entails focusing on some of their aspects and ignoring others. Also in both cases, this selection is shaped by normal commercial pressures and concerns, whether the power of the purse or the need to generate user fees.

The second example concerns something more subtle than the difference between colorful parrot fish and sea grasses. Instead, and again like the case of GM grains, it concerns the ways that these images imply a certain view of the state of the coastal waters and how to secure it, and thereby slight others. The images on those park websites are not simply of colorful fish and coral, they are of individual fish and coral growths. This may seem unexceptionable given the medium of a web page and the nature of underwater photography. However, these images implicitly make "healthy coastal waters" legible in terms of the presence of those individuals, and so help to define ethicality, an ethically desirable coastal environment, in terms of them. That is, healthy coastal waters are those that contain those individuals and unhealthy coastal waters are those that do not. This focus on individuals has important implications.

By focusing on individual fish and coral growths, these images suggest that the pertinent threats to the coastal waters are things that harm those individuals, and that protection of those waters means preventing these threats. For tourist visitors to these waters, this means making sure that the boat you are on uses a mooring buoy rather than an anchor, which can harm coral, and making sure when you are in the water that you do not take a fish or step on a coral. The concern to avoid these threats helps explain the embrace by senior Montego Bay park staff of "catch and release" fishing as a useful venture. With this, they argued, the park could attract fee-paying visitors concerned with the state of the coastal waters. They could fish in protected waters, where fish were supposed to be bigger and more plentiful, and because the fish would be released back into the waters, they would not harm the environment.

Moreover, this focus has implications for how people see not just the coastal waters themselves, but the people associated with them. That is because that focus indicates that those who take a fish or harm a coral are a threat to the health of the coastal waters, unlike those responsible tourists, and need to be stopped. In Jamaica's coastal waters, it is the local, in-shore fishers who take fish and drop anchors that crush coral. The ethicality that these images encourage, then, is implicated in long-running strains that are part of the political economy of those coastal waters, the sort of intermediate context that, Jim Igoe (2010: 382) notes, is routinely absent in representations produced by conservation organizations (for Negril, see Garner 2009). The marine parks in Montego Bay and Negril were developed by those with close ties to the tourism sector (Carrier 2003: 11–18), and park managers shared that sector's long opposition to those local fishers, though this is somewhat more true of those in Montego Bay than it is of those in Negril (Sommer and Carrier 2010). Whether intentionally or not, then, those images and that ethicality are part of a continuing fight over access to, the use of, and, hence, the shaping of a part of the natural environment that is an important economic resource, the coastal waters at Montego Bay and Negril.

There is a further set of implications that arises from those images. In constructing healthy coastal waters in terms of individuals and threats to them, those images and the resulting ethicality ignore populations. There are good practical reasons for this: populations are not easy to portray in the way that individuals or small groups of individuals are; the visitors who are the ultimate source of the revenue that the parks need are attracted by what they can see on a dive or on a cruise in a glass-bottom boat, which excludes abstract things like populations. Good practical reason or not, however, the ignoring of populations in these images and that ethicality have important consequences.

I have described how the focus on individuals implies that certain sorts of activities and certain sorts of people are a danger to the coastal waters. In ignoring populations, these images and that ethicality do something else as well. They ignore the possibility that other sorts of activities and other sorts of people are a danger. Just as populations are abstract concepts, so the threats to them are abstract, but that does not mean that they are immaterial. In Negril and Montego Bay they are material indeed, and here again the ethicality these images encourage is implicated in the political economy of the coastal waters. That is because the activities at issue are those of the tourism sector, and the people are the tourists the sector seeks to attract. Both Montego Bay and Negril live off that tourism.

In the forty years before I studied it, Montego Bay grew from a port town of about 45,000 to a commercial and tourist city with a population in the metropolitan area of about 100,000. It is the largest tourist destination on the island and its airport handled three-quarters of the 1.35 million tourists entering the country in 2003 (Bakker and Phillip 2005: fig. 18). Negril is about 50 miles (80 km) to the west of Montego Bay at the western tip of Jamaica. In that same forty years it went from being little more than the site of a fishing camp to a town of about 20,000, becoming the country's third largest tourist destination, and containing almost a quarter of Jamaica's tourist accommodation. In 2003 the place attracted about 275,000 visitors and its hotels employed almost 7,800 people (Bakker and Phillip 2005). The growth of the population of these places, the result of their success as tourist destinations, placed strains on the coastal ecosystems, made worse by the fact that basic municipal services like sewerage and rubbish collection were not increased to keep pace. Further, tourists place exceptional demands on these services: in 1994, three-fifths of the hotel waste-water in Jamaica was either untreated or treated inadequately (Burke 2005: 11), and the average tourist in Jamaica produces almost four times as much solid waste daily as the average resident (Thomas-Hope and Jardine-Comrie 2005: 3). These land-based activities are the main environmental threat to the coastal waters that these parks oversee.

Tourists, however, do not seem to worry about this, but instead worry about those Jamaican in-shore fishers. The former head of the Negril park said that hotel managers would tell her that guests were complaining about fishing boats that were in park waters. Even though fishers had the right to be in those waters so long as they were not fishing in prohibited zones, the park's head needed the goodwill of the hotels, and so was under pressure to ban those boats from substantial parts of park waters. Tourists did not, however, appear to complain about the more substantial damage to the coastal environment caused by their own activities and the hotels where they were staying. In focusing on their encounter with Jamaica's coastal waters while ignoring the relationships and institutions involved in generating that encounter, those tourists of course illustrate Scott's (1998) observation that one aspect of Modernity is the tendency to see things independently of their context, which itself echoes what Marx said about the fetishism of commodities (see Carrier 2010).

My point about the focus on individuals in the images on park websites again echoes one that I made about GM grains and world hunger: complex sets of activities and relationships are radically simplified, stressing one way of understanding the world and slighting others. In the case of GM grains, everything from the seed going into the ground to food in people's stomachs was slighted in favor of the constitution of the seed. In the case of Jamaica's coastal waters, the nature and consequences of everything going on in regional ecosystems is slighted in favor of the presence of colorful fish and coral. The constitution of the seed is significant, as is the presence of the fish and coral head. However, in focusing attention on these, other things are slighted, with the result that the problems of world hunger and the health of Jamaica's coastal waters are framed in certain ways rather than others. And just as the framing that is associated with the advocacy of GM grains illustrates the power of the purse, so does the framing that emerges from those images on park websites.

That purse, however, is not so straightforward as it is with GM seeds: the Montego Bay Marine Park Trust and the Negril Coral Reef Preservation Society are no powerful corporations. Rather, a significant part of that purse is under the control of individual tourists. In aggregate they generate the user fees that can allow those parks to survive, which is the reason for the images on the websites. The other part of that purse is in the hands of the companies that make up the Jamaican tourism sector, and those within it say that tourists want the sparkling clear waters and striking fish and coral that the parks' websites portray. Although these companies are no match for Monsanto, they operate in a much smaller political-economic realm, and within that realm they are powerful, both because of the size of their purse and because, since the country underwent structural adjustment in the 1980s (Bartilow 1997; Bloom et al. 2001), the Jamaican economy has relied on tourism to survive and the Jamaican government has supported the tourism sector.

The companies in that sector, and the Jamaican government, want coastal waters that will make Montego Bay and Negril attractive to tourists. Although their goals are different from the more modest desire of the two marine parks to generate user fees, they are like the parks in wanting appealing images of what a tourist might see in those coastal waters. All three, then, have simple commercial reasons to focus on colorful fish and coral, and all three end up encouraging the definition of "healthy environment" pretty much as one that attracts tourists (Sommer and Carrier 2010: 190–191).

This case of marine parks in Jamaica complements what I said about GM seeds, that different ways of approaching problems in the world entail different definitions of, and stress different aspects of, what that problem is. In the concrete case of the Negril and Montego Bay parks, I used the idea of ethicality to illustrate this process in more detail. I also showed how commercial interests very different from those I sketched in the GM case can be important in that process. In the next section I extend and complicate that argument, by considering some of the implications of the point that those images on park websites are intended to attract and persuade.

The Power of the Image

The people who view these Internet images are not likely to know much of coastal tropical ecosystems, any more than the viewer of images of a car in an advertisement is likely to know much of automobile design and manufacture. This does not mean that these viewers are tabula rasa, happily accepting and adopting those images and their implications. Rather, they come with a set of presuppositions and values, a cultural background that shapes their interpretation of those images, and thus the resulting ethicality. If those who decide what images to present are wise, they will take that background into account when they make their selection. If they are unwise, the images are unlikely to achieve the desired result and another selection will have to be made if the enterprise is to get the money that it needs.

Even though it is not directly about understandings of the natural world, I illustrate the importance of that cultural background with what Peter Luetchford (2012) says about images associated with a different sort of ethical concern, coffee that is produced in ways that are not exploitative. Like "healthy coastal waters," "nonexploitative coffee" is an abstraction, and the images that are associated with it provide fairly concrete instances of that abstraction and so come to define it. Luetchford is concerned with the images on bags of "relationship coffees," which portray the growers and, thereby, seek to combine grower and coffee into a single ethical product. If this coffee is to sell, the images need to appeal to people in supermarket aisles. The sorts of people portrayed as growers in those images are self-reliant smallholders, working their own plots of land and tending their own trees. Such people are appealing to many of the Americans and Europeans who see those images. However, just as the images on those park websites select from, and hence simplify, a more complex reality, so too do those pictures of growers.

This simplification is clear from what Luetchford describes of the more complex reality of coffee production. The cooperatives he studied in Costa Rica that produce coffee certified as Fairtrade did have members who were self-reliant smallholders, but such people were not that common. Some growers were landless sharecroppers, a number used migrant workers from Nicaragua during the harvest, some were substantial landowners who devoted most of their land to cattle and set aside only a small part for coffee, and so on. Given what Luetchford describes, an image of a prosperous cattle rancher or half a dozen Nicaraguan migrant workers living in barracks would represent the production of Fairtrade coffee in that area as accurately as would the image of the sturdy smallholders. However, neither the rancher nor the Nicaraguans accord with the orientations and presuppositions of those the coffee company seeks to attract; their images are not used; the ethicality of nonexploitative coffee is shaped in certain ways rather than others.

Luetchford's description of the images on bags of Fairtrade coffee complicates what I said in the two preceding sections about the representations that people confront and the way that those representations shape the way that people understand aspects of their world. People who perceive those representations interpret them in light of their values and presuppositions; those who project those representations select them in light of how they think they will be interpreted by the people they seek to attract. These complications are pertinent if we want to understand the emergence and content of ethicalities in particular, and more generally the ways that commercial pressures shape people's understandings of their surroundings. They indicate that the pertinent values and presuppositions need not in any obvious way be about the substance of what is portrayed in those images or the abstractions that they are supposed to represent.

I mentioned that representations are likely to be selected in light of suppositions about the sorts of people that those who project them want to attract. Not every potential viewer is, after all, equally desirable, and that desirability can itself be complex. What Kathy Rettie (2009) says of Parks Canada, the Canadian national park authority, illustrates this. Like those marine parks in Jamaica, Canada's parks need the visitors (Eagles 2002), and they seek to attract visitors who will approach the national parks in a way that the authority wants, as preserving Canada's distinctive natural endowment and its past. Such visitors will walk the trails and view the interpretative displays and reenactments that present the place and its wildlife, as well as the indigenous people and early settlers who lived there and the ways that they used and appreciated what they found there. As distinct from such people, Rettie says, the parks do not want visitors who seek only to eat in the restaurants and drink in the bars, even though such visitors erode no land, disturb no wildlife, never stray from the paved roads. The images these parks produce are selected, and Canada's natural heritage is made legible, appropriately.

It is likely that the sort of concerned Canadian lover of nature that the authority wants would not find those partygoers any more attractive than Parks Canada does. In associating certain sorts of activities with certain sorts of people, those concerned Canadians exhibit what Jean Baudrillard (1981) describes as assessing things (like going for a drink or a meal in a park restaurant) not simply in terms of the desirability of those things themselves, but also in terms of the desirability of those associated with them (like the sort of people who go to the park only for a drink or a meal). Such assessments may reflect the social distinction that concerned Bourdieu (1984) more than they reflect, say, the elements of Jamaica's coastal waters and their interactions. Even so, because they are likely to affect the ways that those waters are represented, they will affect the ways that aspects of them are made legible. This will shape the ethicality of "healthy ecosystem", and in turn shape the ways that people deal with those waters, and so shape the waters themselves.

A less obvious instance of that shaping, and of its links with social distinction, is found among those interested in authentic foods, produced in traditional ways that respect and reflect the environment. Many different people value such food, and they do so for many different reasons, but social distinction can be an important aspect of these reasons (see, e.g., West 2010). The Slow Food movement is an example of this, for it espouses such respect for local environments and the people who produce food there and is, at the same time, associated with discriminating tastes and the sort of people who can afford to cultivate them. That movement is another instance of ethical consumption commerce, and as Cristina Grasseni (2012) shows, the pressures of that commerce influence the ways that local environments and producers are understood, while also shaping the ways that those producers act and, thereby, influence that environment.

Conclusion

As the other articles in this special issue describe, there are many aspects of the relationship of capitalism and nature, and many ways to approach it. My purpose here, however, has been a simple one. It is to focus on commercial and economic processes of the middle range, in order to show how the routine and unexceptionable aspects of honest commercial practice in capitalist systems can affect the ways that people understand their world, including nature and the natural environment. As I have described them, the companies that make GMO seeds, the marine parks in Jamaica and the companies in the tourism sector there, those who grow and those who sell Fairtrade coffee, have been mundane and unexceptionable in their orientations and activities. In this article I have attended to that mundane and unexceptionable, to show how what goes on there shapes the ways that people understand their world.

Although my approach relies on a fairly simple analytical frame, the implications that I have drawn from that frame make at least certain aspects of what we study appear more complex. Those implications arise from the way that the straightforward operations of ethical consumption commerce affect ethicality, the ways that people understand abstract notions like "healthy coastal waters." As I have argued, the way such commerce works means that those understandings are likely to be shaped by commercial considerations, however unrecognized, unintended, or even unconscious that shaping may be. At the same time, I have shown how those shapings themselves are likely to be more complex than they appear at first glance. That is so because the information and especially the images that people receive are not absorbed naively. Rather, they are likely to be interpreted in complex ways that reflect concerns and interests that need have little or nothing to do with their overt content. Moreover, because that information and those images are intended to serve a commercial purpose, there is pressure on those who produce them to take those concerns and interests into account.

Although the fundamental processes described in this article are straightforward, they are one aspect of the relationship between capitalism and nature that can have important and unexpected effects on the ways that people understand the world around them, and hence on the ways that they act on and shape that world.

■ **JAMES G. CARRIER** has taught and done research on aspects of economy in Papua New Guinea, the United States, and Great Britain, as well as studying environmental conservation in Jamaica. He is Hon Research Associate in anthropology at Oxford Brookes University, adjunct professor of anthropology at the University of Indiana, and Associate at the Max Planck Institute for Social Anthropology. His publications include the edited volumes *Meanings of the Market* (1997), *Virtualism, Governance and Practice* (2009, with P. West) and *Ethical Consumption* (2012, with P. Luetchford).

■ REFERENCES

African Post Harvest Losses Information System. n.d. http://www.aphlis.net/index.php?form=home.

Applbaum, Kalman. 2006. "Educating for Global Mental Health: American Pharmaceutical Companies and the Adoption of SSRIs in Japan." Pp. 85–110 in *Pharmaceuticals and Globalization: Ethics, Markets, Practices*, ed. A. Petryna, A. Lakoff and A. Kleinman. Durham, NC: Duke University Press.

Applbaum, Kalman. 2009. "Getting to Yes: Corporate Power and the Creation of a Psychopharmaceutical Blockbuster." *Culture, Medicine and Psychiatry* 33, no. 2: 185–215.

Argyrou, Vassos. 2005. *The Logic of Environmentalism: Anthropology, Ecology and Postcoloniality.* Oxford: Berghahn Books.

Bakker, M., and S. Phillip. 2005. *Travel and Tourism—Jamaica—February 2005.* London: Mintel International Group.

Bartilow, Horace A. 1997. *The Debt Dilemma: IMF Negotiations in Jamaica, Grenada and Guyana.* London: Macmillan.

Baudrillard, Jean. 1981. *For a Critique of the Political Economy of the Sign.* St. Louis, MO: Telos Press.

Bloom, David E., Ajay Mahal, Damien King, Aldrie Henry-Lee, and Philip Castillo. 2001. *Occasional Paper: Globalization, Liberalization and Sustainable Human Development: Progress and Challenges in Jamaica.* Geneva: United Nations Conference on Trade and Development, and United Nations Development Programme.

Bourdieu, Pierre. 1984. *Distinction: A Social Critique of the Judgement of Taste.* London: Routledge & Kegan Paul.

Burke, R.I. 2005. *Environment and Tourism: Examining the Relationship Between Tourism and the Environment in Barbados and St. Lucia.* Neuilly-sur-Seine: PriceWaterhouseCoopers.

Carrier, James G. 2001. "Social Aspects of Abstraction." *Social Anthropology* 9, no. 3: 243–256.

Carrier, James G.. 2003. "Mind, Gaze and Engagement: Understanding the Environment." *Journal of Material Culture* 8, no. 1: 5–23.

Carrier, James G.. 2010. "Protecting the Environment the Natural Way: Ethical Consumption and Commodity Fetishism. *Antipode* 42, no. 3 (Special Issue): 668–685.

Carrier, James G.. 2012. "Introduction." Pp. 1–35 in *Ethical Consumption: Social Value and Economic Practice*, ed. J.G. Carrier and Peter G. Luetchford. Oxford: Berghahn Books.

Eagles, Paul F.J. 2002. "Trends in Park Tourism: Economics, Finance and Management." *Journal of Sustainable Tourism* 10, no. 2: 132–153.

Edgerton, David. 2008. *The Shock of the Old: Technology and Global History since 1900.* London: Profile Books.

Eichenwald, Kurt. 1995. *Serpent on the Rock.* New York: HarperCollins.

Emmott, Bill. 2008. "GM Crops Can Save Us from Food Shortages." *Telegraph*, 17 April. www.telegraph .co.uk/comment/3557344/GM-crops-can-save-us-from-food-shortages.html (accessed 1 Sept. 2012)

Errington, Frederick, and Deborah Gewertz. 2001. "On the Generification of Culture: From Blow Fish to Melanesian." *Journal of the Royal Anthropological Institute* (NS) 7, no. 3: 509–525.

Escobar, Arturo. 1999. "After Nature: Steps to an Antiessentialist Political Ecology." *Current Anthropology* 40, no. 1: 1–30.

Food and Agriculture Organization. 1989. *Prevention of Post-Harvest Food Losses: Fruits, Vegetables and Root Crops.* Rome.

Garner, Andrew. 2009. "Uncivil Society: Local Stakeholders and Environmental Protection in Jamaica." Pp. 134–154 in *Virtualism, Governance and Practice: Vision and Execution in Environmental Conservation,* ed. James G. Carrier and Paige West. Oxford: Berghahn Books.

Grasseni, Cristina. 2012. "Re-inventing Food: Beyond Consumption?" Pp. 198–216 in *Ethical Consumption: Social Value and Economic Practice,* ed. James G. Carrier and Peter G. Luetchford. Oxford: Berghahn Books.

Hillel, Daniel. 1997. *Small-Scale Irrigation for Arid Zones: Principles and Options.* Rome: Food and Agriculture Organization.

Ho, Karen. 2009. *Liquidated: An Ethnography of Wall Street.* Durham, NC: Duke University Press.

Igoe, Jim. 2010. "The Spectacle of Nature in the Global Economy of Appearances: Anthropological Engagements with the Spectacular Mediations of Transnational Conservation." *Critique of Anthropology* 30, no. 4: 375–397.

Ingold, Tim. 1988. "Tools, Minds and Machines: An Excursion in the Philosophy of Technology." *Techniques et Culture* 12: 151–176.

Lépinay, Vincent. 2011. *Codes of Finance: Engineering Derivatives in a Global Bank.* Princeton, NJ: Princeton University Press.

Luetchford, Peter G. 2012. "Consuming Producers: Fair Trade and Small Farmers." Pp. 60–80 in *Ethical Consumption: Social Value and Economic Practice,* ed. James G. Carrier and P.G. Luetchford. Oxford: Berghahn Books.

Lukes, Steven. 1974. *Power: A Radical View.* London: Macmillan.

Merton, Robert K. 1968. "On Sociological Theories of the Middle Range." Pp. 39–72 in R.K. Merton, *Social Theory and Social Structure.* Enlarged edition. New York: Free Press.

Moberg, Mark, and Sarah Lyon. 2010. "What's Fair? The Paradox of Seeking Justice Through Markets." Pp. 1–24 in *Fair Trade and Social Justice: Global Ethnographies,* ed. S. Lyon and M. Moberg. New York: New York University Press.

Muniesa, Fabian. 2007. "Market Technologies and the Pragmatics of Prices." *Economy and Society* 36, no. 3: 377–395.

Ouroussoff, Alexandra. 2010. *Wall Street at War: The Secret Struggle for the Global Economy.* Cambridge: Polity Press.

Rettie, Kathy. 2009. "A Culture of Conservation: Shaping the Human Element in National Parks." Pp. 66–83 in *Virtualism, Governance and Practice: Vision and Execution in Environmental Conservation,* ed. James G. Carrier and Paige West. Oxford: Berghahn Books.

Scott, James C. 1998. *Seeing Like a State.* New Haven: Yale University Press.

Sen, Amartya. 1981. *Poverty and Famines: An Essay on Entitlement and Deprivation.* Oxford: Clarendon Press.

Sommer, Gunilla, and James G. Carrier. 2010. "Tourism and its Images of Tourists, Traders and Fishers in Jamaica." Pp. 174–196 in *Tourism, Power and Culture: Anthropological Insights,* ed. Donald Macleod and J.G. Carrier. Bristol: Channel View.

Stone, Glenn Davis. 2010. "The Anthropology of Genetically Modified Crops." *Annual Review of Anthropology* 39: 381–400.

Stone, Glenn Davis, and Dominic Glover. 2011. "Genetically Modified Crops and the 'Food Crisis': Discourse and Material Impacts." *Development in Practice* 21, nos. 4–5: 509–516.

Thomas-Hope, E., and A. Jardine-Comrie. 2005. "Valuation of Environmental Resources for Tourism: The Case of Jamaica." IRFD World Forum on Small Island Developing States. http://irfd.org/events/wfsids/virtual/papers/sids_ethomashope.pdf (accessed 22 December 2011).

West, Paige. 2010. "Making the Market: Specialty Coffee, Generational Pitches, and Papua New Guinea." *Antipode* 42, no. 3 (special issue): 690–718.

Zaloom, Caitlin. 2006. *Out of the Pits: Traders and Technology from Chicago to London.* Chicago: University of Chicago Press.

Neoliberalism and the Production of Environmental Knowledge

Rebecca Lave

▪ ABSTRACT: In order for nature/society scholars to understand the dynamics of environmental appropriation, commercialization, and privatization, we must attend to the production of the environmental science that enables them. Case studies from anthropology, geography, history of science, science and technology studies, and sociology demonstrate that the neoliberal forces whose *application* we study and contest are also changing the *production* of environmental knowledge claims both inside and outside the university. Neoliberalism's core epistemological claim about the market's superiority as information processor has made restructuring the university a surprisingly central project. Further, because knowledge has become a key site of capital accumulation, the transformative reach of neoliberal science regimes extends outside the university into the various forms of extramural science, such as citizen science, crowdsourcing, indigenous knowledge, and local knowledge. Neoliberal science regimes' impacts on these forms of extramural science are strikingly similar, and quite different from the most common consequences within academia.

▪ KEYWORDS: neoliberal university, citizen science, crowdsourcing, indigenous knowledge, local knowledge, science regimes

Political ecologists and critical nature/society scholars have a long history of studying the application of environmental physical science to promote colonial and neocolonial agendas (e.g., Blaikie 1985; Davis 2007; Fairhead and Leach 1996; Leach and Mearns 1996; Neumann 1988; Turner 1999), legitimize state or corporate appropriation of local resources (Braun 2002; Hecht 1985; Hollander 2008; McCarthy 2002; Prudham 2005; Sayre 2002), and deny environmental and biophysical damage caused by corporations (Guthman and Dupuis 2006; Kirsch 2011), By contrast, we have paid comparatively little attention to the production of environmental knowledge claims (but see Demeritt 1998, 2001, 2006; Duvall 2011; Ellis and Waterton 2004; Forsyth 2003; Raffles 2002; Robertson 2006) even though we cannot understand the environmental management frameworks and policies applied at our field sites without analyzing the knowledge claims that enable them.

The production of those claims is a surprisingly central site of neoliberalization. While there is much research still to be done on the neoliberalization of universities, doing so is relatively straightforward, and there is already a substantial body of literature about it. Knowledge claims produced outside the academy, which have taken on increasing economic and academic significance since 1980, present a far less centralized unit of analysis (indigenous knowledge, citizen science, crowdsourcing, etc.). Yet neoliberal impacts on nonacademic knowledge production

Environment and Society: Advances in Research 3 (2012): 19–38 © Berghahn Books
doi:10.3167/ares.2012.030103

turn out to be strikingly similar across this disparate range of settings, and also notably different from the most common consequences of the neoliberalization of academia.

Science Regimes

In political ecology and critical nature/society studies more broadly, we attend carefully to the interested application of environmental science in the service of resource appropriation, commodification, and privatization. But the political-economic forces we study in the field have an equally powerful impact on the ways that environmental science is produced in the first place, shaping a) the questions investigated and, perhaps more important, funded; b) the beneficiaries of science; and c) the "principles of vision and division" with which natural scientists (and we, too) think (Bourdieu 1998). We are accustomed to acknowledging political-economic influences on the production of biomedical knowledge claims. We shake our heads over reports of conflicts of interest in clinical drug trials (Vioxx, anyone?), but somehow lose sight of the fact that quite similar dynamics are at work in environmental physical science as well.

To more fully incorporate these dynamics into our research, we need an analytical model that links scientific production, circulation, and application to each other and to larger political economic forces. Conventional wisdom paints a simple, unidirectional picture of this process (Figure 1; Goldman and Turner 2011). In this model, knowledge is produced by scientists and then transferred to people who apply that knowledge in the way scientists envisioned. There are very few actors, no feedback among the stages, and no agency on the part of those who apply scientific knowledge. Nor is there any recognition of the broader political-economic context within which this process plays out.

Figure 1. Linear model of knowledge transmission

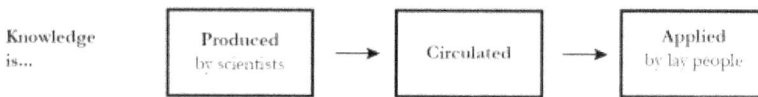

The typical political ecology version of this model is somewhat more nuanced in that the tension between individuals' agency and political economic forces is clearly visible (Figure 2). Production is typically black boxed as an unproblematic feeder into more important processes, though, and there is still no feedback among the stages of the model.

Figure 2. Typical political ecology model of knowledge transmission

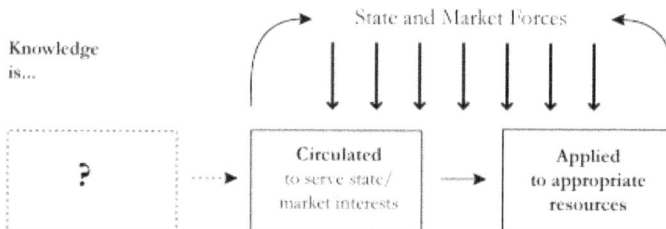

Scholarship in science and technology studies (STS) paints a far more complicated picture, demonstrating the deep interconnections among knowledge production, circulation, and application (Figure 3). Historian of science Dominic Pestre (2003) coined the term *science regime* to describe this complexity, and the tensions between the agency of individuals and the larger political-economic context in which they are embedded. Pestre's research demonstrates that the source and guiding philosophy of science funding and management at a particular place and time deeply shapes scientists' conditions of production, the content they produce, and how that content is circulated and applied. Scientists have never worked under circumstances of their own choosing, and those circumstances shape (but do not determine) their research practice and even their findings.

Figure 3. Model of science regimes

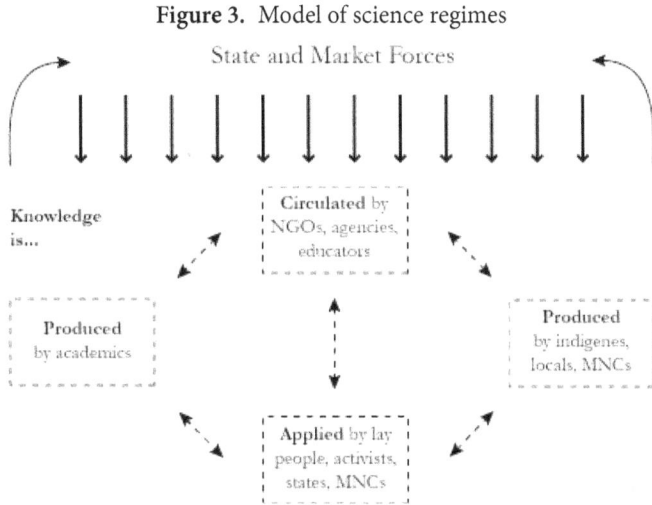

State and Market Forces

Knowledge is...

Circulated by NGOs, agencies, educators

Produced by academics

Produced by indigenes, locals, MNCs

Applied by lay people, activists, states, MNCs

In any given historical moment and institutional context, relations among scientists, states, and economic elites create a distinctive science regime. In the United States, European Union, and other parts of the world—including Australia, China, New Zealand, and Japan—the current science regime is increasingly neoliberal.

Neoliberal Science Regimes and Their Impacts on Academic Knowledge Production

Universities are embroiled in a messy, uneven, and contested process of neoliberalization. They have lots of company: the past three decades have brought the neoliberalization of many formerly public projects, such as K-12 education, urban planning, health care, and environmental management. As demonstrated by Phillip Mirowski (2011), however, *science and its current institutional locus—the university—turn out to be surprisingly central to neoliberal agendas.* Beyond the arguments about market-based efficiency that can be applied to all of the sectors listed above, the keystone of neoliberal philosophy is an epistemological argument that the market is the superior information processor, knowing more than any individual ever can. Thus in neoliberal thought and increasingly in neoliberal science policy, university professors and researchers are viewed as at best embarrassingly misguided in their truth claims, and at worst actively harmful to the proper functioning of society. Universities have thus been subject to passionate attacks from neoliberal theorists for decades, as evidenced in the writing of Friedrich von Hayek and other members of the Mont Pelerín Society, such as Milton Friedman and George Stigler.

The policies and practices that stem from neoliberalism's very distinctive set of claims about epistemology and the university take forms that are nationally specific and operationalized differently by discipline and even by campus. As Canaan and Shumar describe it: "the context of the political economy of globalization and the rise of neoliberal economic ideology … shape the stories of higher education being told in different countries and under different types of educational systems. … [These] local differences … are the result of different penetrations of media, migration, cultural, economic, finance and other flows" (2008: 1). Despite this variation, there are commonalities. I address five here, synthesized from case studies on a wide range of academic fields and national contexts: reduction in public funding for universities, separation of teaching and research, the replacement of peer review with market-based mechanisms, the tyranny of relevance, and the formidable strengthening of intellectual property protections.

The first common characteristic of neoliberal science regimes is redistribution of the costs of higher education from the public sector to the private (Mirowski 2011; Nedeva and Boden 2006). This trend is strikingly broad, with substantive drops in the percentage of federal funding for universities in almost all industrialized countries. A 2007 report by the Institute for Higher Education Policy found that between 1995 and 2003 the proportion of university funding provided by the public sector dropped in all but 4 of 51 countries surveyed and was replaced by private funding, nearly two-thirds of which came from household contributions (Hahn 2007: 4–6). Vincent-Lancrin (2006) describes a similar trend: public funding for universities (excluding grants for specific research projects) in the 16 countries of the Organization for Economic Cooperation and Development he surveyed declined from 78 percent to 65 percent between 1981 and 2003, with Australia, Denmark, Finland, Greece, Ireland, New Zealand, Spain, Turkey, and the UK all experiencing larger than average drops. Further, the distribution of public funds for universities within nation states was increasingly determined through research evaluation metrics (Vincent-Lancrin 2006: 179–182). Thus the actual decrease in public funding has been more severe for many universities than the national average would suggest, while a few elite research institutions have enjoyed funding increases.

In the United States, state-level funds for public universities have fluctuated substantially over the past decade. Large cuts in the early 2000s (Mirowski 2011; Rizzo 2004) were followed by modest increases in the middle of the decade (Slaughter and Rhoades 2004), now undone by the severe economic downturn that began in 2007. According to the Center on Budget and Policy Priorities, in 2011 43 out of 50 US states cut spending for higher education, in many cases quite substantially;[1] nationwide, per student public spending has dropped to a 25-year low (Martin and Lehren 2012).

Ideologically, this striking downward trend in public funding for universities has been driven by the reframing of education as an individual's investment in her own human capital rather than a public investment for the greater good of society (Kaye et al. 2006; Lambert et al. 2007; Mirowski 2011; Nedeva and Boden 2006; Slaughter and Rhoades 2004). As Slaughter and Rhoades explain, the shift in federal funding for higher education in the United States from direct funds to universities to loans to individual students (or their parents) has been justified by,

> the growing idea that higher education is largely a private good, with the benefits going primarily to individual students (who are increasing their human capital), so students and their families should be expected to bear a larger share of the costs. The externalities of higher education, the social benefits beyond the student, are overlooked and undervalued. Over time, the benefits of any expanded public investment in broader access to higher education have come into question. (2004: 283)

A second common characteristic of neoliberal science regimes is the increasing separation of research and teaching (Gibbons et al. 1994; Lambert et al. 2007; Mirowski 2011; Nowotny et al. 2001; Slaughter and Rhoades 2004), with the former increasingly privileged as a source of external revenue. In the US, the reduced emphasis on teaching has taken the form of differentiation among academic staff. The percentage of untenured faculty—both itinerant and long-term adjunct positions—has skyrocketed. The American Association of University Professors' Contingent Faculty Index, a study of hiring practices at 2,617 US colleges and universities, found that between the 1970s and 2005 the percentage of tenure and tenure-track faculty dropped sharply from approximately 57 percent to just 35 percent (reported in Gravois 2006). In Europe the separation of research and teaching is occurring at the campus level. As mentioned above, performance-related metrics like the UK's Research Assessment Exercise redistribute funding among campuses, cutting funds and increasing course loads for the majority of universities (Nedeva and Boden 2006; Strathern 2000).

A third common characteristic of neoliberal science regimes is the circumvention of existing peer-review systems. Sergio Sismondo (2007, 2009) has demonstrated that approximately 40 percent of all articles in biomedical journals are planned and written by biomedical companies in a comprehensive privatization that he refers to as "ghost management." In the most egregious case so far uncovered, Elsevier produced multiple issues of six fake journals. These journals appeared to be peer reviewed, did not disclose their corporate funding (Grant 2009), and were distributed to tens of thousands of physicians in Australia between 2000 and 2003 (Singer 2009).

Beyond such illicit circumventions of peer-review gatekeeping, there are now explicitly privatized and commercialized peer-review systems, such as the Faculty of 1000 (F1000), an online site in which nominated faculty provide a brief description and rating of published articles. F1000 seems to share some weaknesses of the current peer-review system, such as cronyism and uncredited delegation: the *Chronicle of Higher Education* reported that many of the nominated members are assigning these unpaid reviews to their students rather than writing them themselves (Macilwain 2011). The larger issue, however, is that F1000 is now selling its privately produced reviews and research quality evaluations as an alternative to existing metrics through a partnership with EBSCO, one of the largest pay-per-view online academic content providers.[2]

A fourth common characteristic of neoliberal science regimes is the tyranny of relevance: the prioritization of knowledge produced to meet market needs at the expense of noncommercial research in the humanities, much of the social sciences, and even basic science (Canaan and Shumar 2008; Gibbons et al. 1994; Kleinman 2003; Lave et al. 2010b; Moore et al. 2011; Nedeva and Boden 2006). Canaan and Shumar argue that:

> the emphasis today is on applied research that can be turned into a marketable commodity. … [As a result,] *the humanities and the social sciences are becoming increasingly ghettoized* … [viewed by administrators as] a necessary sign of university education that should either be rationalized by making the teaching cheaper (such as online courses in the U.S. or by giving universities less income for arts and social sciences than for business, law and natural sciences in the U.K.) or should be instrumentalized by somehow bringing this research in line with the more profitable forms of research at the university. (2008: 10 and 16; emphasis added)

This trend was already clear in the early 1990s, as documented in *The New Production of Knowledge*, a germinal work on the transformation of science in Europe and the United States by Gibbons et al. (1994). They wrote that, "Less and less it [research] is curiosity-driven and funded out of general budgets which higher education is free to spend as it likes; more and more it is in the form of specific programmes funded by external agencies for defined purposes," as "the scientific industrial system [attempts] *to filter science through the sieve of industrial needs*" (Gibbons et al. 1994: 78, 163; emphasis added).

In addition to narrowing the topics of research, the neoliberal emphasis on market relevance leads many researchers to jump from topic to topic in search of funding rather than pursing a sustained, self-directed research program (Gibbons et al. 1994: 86), creating a profound instrumentalization of academic research. According to Nedeva and Boden, under neoliberal science regimes,

> academics produce what they can sell and what is immediately and directly useable by "customers." … There is an observable epistemic shift whereby academic[s] research in areas which generate financial support, they generate knowledge that they can sell and tend to present it as immediately and directly useful by "customers." (2006: 278, 279–280)

The final common characteristic of neoliberal science regimes' impacts within universities stems from the stunning expansion of intellectual property protections, which has been central to neoliberal strategies of capital accumulation (Mirowski 2011; Nedeva and Boden 2006; Nowotny 2005; Nowotny et al. 2001; Tyfield 2010). As Pestre notes, the common way to characterize that expansion, "is to speak of a new movement of enclosure. The analogy is that we face a privatization of the 'commons of the mind' (what public science used to be) which recapitulates, several centuries later, the privatization of the 'common land' in early modern Britain" (2005: 34–35).

STS scholars have documented this new wave of enclosure in fascinating detail through the recent history of patent expansion in the United States (Biagioli 2006; Coriat and Orsi 2002; Geiger and Sa 2008; Mirowski 2011; Popp-Berman 2008). Briefly, beginning in 1980 with *Diamond vs. Chakrabarty* and the Bayh-Dole Act (and continuing on through further legal cases, acts of Congress, and executive orders by US presidents), US law has strikingly expanded what can be given intellectual property protection through patenting. It is now possible to patent anything from living beings to business practices, and to patent discoveries stemming from federally funded research whether by corporations, which receive the lion's share of federal research dollars, or by researchers at universities. The dramatically fortified US patent regime was then extended internationally in the form of the Trade Related Aspects of Intellectual Property Rights Agreements through a small coterie of US corporations' successful hijacking of the Uruguay Round of the General Agreement on Tariffs and Trade in 1994 (see Tyfield 2010 for a very useful history).[3]

The treatment of *knowledge as a target of appropriation,* an undercapitalized realm that can restart the process of capital accumulation, is a signature of neoliberal science regimes (Tyfield 2010). As Canaan and Shumar write, "In the new economy, knowledge is a critical raw material to be mined and extracted from any unprotected site; patented, copyrighted, trademarked or held as a trade secret; then sold in the marketplace for a profit" (2008: 4). Academia, however, is only one source of this raw material. Knowledge produced outside the university has also been central to neoliberal strategies of accumulation, and thus to neoliberal science regimes. Before moving on to extramural science, it is worth noting that the environmental sciences—particularly newly prominent ones like the study of climate change, biodiversity loss, or environmental mutagens—seem to be especially vulnerable to these last two trends. Research in these fields has been catalyzed by a sense of crisis rather than by scientific breakthroughs, resulting in relatively underdeveloped content often cobbled together from pieces of preexisting fields. In addition, these fields emphasize the complexity and particularity of the systems they study, leading to high levels of uncertainty. These new environmental sciences focus on issues in which the general population and markets (in cases such as carbon trading and weather derivatives) have powerful interests, which can lead them to intervene in scientific debates. Finally, unlike say particle physics, many environmental sciences focus on subjects about which lay people have substantial knowledge. All of this makes the environmental sciences more open both to the neoliberal emphasis on privatized and commercialized knowledge, and to extramural knowledge providers (Lave 2012b).

Neoliberal Science Regimes' Impacts Outside the Academy

Universities are the most widely recognized sources of knowledge production, but they are hardly the only source as phenomena such as the resurgence of bioprospecting demonstrate. Following the lead of neoliberal science and accumulation regimes beyond the ivory tower to new sources of intellectual raw material turns out to be far harder than it initially appears. Unlike the university, which presents a broad but relatively well-defined target, extramural knowledge production[4] is geographically diffuse and confusingly demarcated. Instead of the fairly integrated body of existing literature on the university reviewed above, scholarly research on extramural knowledge production is fragmented into separate literatures on amateur science, indigenous knowledge, local knowledge, crowdsourcing, commercial science, and citizen science. These literatures rarely refer to each other despite the fact that most of them are studied in a common set of fields (anthropology, geography, history of science, sociology, and STS); puzzlingly, they do not even seem to view each other as engaged with similar issues. Further, most of the literatures on extramural science do not address neoliberalism despite the fact that the resurgence of academic and policy interest in extramural science has very clear neoliberal roots. In this section I pull together the disconnected literatures on extramural knowledge production, highlighting their commonalities and demonstrating that neoliberal science regimes are having notably similar impacts on them.

History and Forms of Extramural Knowledge Production

As late as the mid-1800s, there was far less separation between full-time scientists and those we would today regard as amateurs. Certainly, there were very real class differences[5] between those who could afford to devote their energy to science and those for whom it could only be a part-time pursuit, but they saw themselves as engaged in a common enterprise. Part- and full-time scientists pored over the same texts, shared their knowledge and collections through local societies, and conveyed their findings in periodicals that did not screen authors by professional status (D. Allen 1976; Keeney 1992; Knell 2000; Kohler 2006; Oleson and Brown 1976; Secord 1994). Full-time scientists needed a network of collectors to bring them specimens, and some amateurs relied on professionals for financial support and in-depth information (Keeney 1992; Reingold 1976).

In the mid- to late-1800s, full-time scientists began a process of *professionalization* that excluded their former colleagues. This process featured certification through formal university degrees in science, and the foundation of both professional societies to which only full-time scientists could belong, and of journals for which only professionals could write. The shift from field science to lab science was also significant, as the expensive apparatus required for lab science shut amateurs out of mainstream scientific practice (Keeney 1992; Reingold 1976). The relocation of scientific practice from venues open across class, such as the pubs that were the center of amateur botany in England, to venues open only to the upper classes was another powerful technique of exclusion (Secord 1994).

Professionalization was accompanied by *appropriation*. The collectively developed body of knowledge was recast as the product and property of white, Western, professional scientists who both published without credit to the broader community that enabled their conclusions, and limited public access to still growing collections (Secord 1994). By the late 1800s, only professionals had legitimate access to the domain of science (Reingold 1976).

Extramural science did not dissolve in the face of these twin processes of professionalization and appropriation, but it certainly dropped from academic view. Neither natural nor social scientists paid much attention to knowledge generated outside the academy during most of the twentieth century. Around 1980, however, extramural science suddenly experienced a dramatic resurgence in visibility in corporate, national, and international policy. Pharmaceutical companies embarked on the current wave of bioprospecting/biopiracy in the 1980s (Brush and Stabinsky 1996; Hayden 2003; Shiva 2001), and the Environmental Justice movement began to gain traction in places like Woburn, Massachusetts (Brown and Mikkelsen 1990) and Cancer Alley in Louisiana (B. Allen 2003). Canada required integration of indigenous knowledge into environmental management policies in 1985 (Nadasdy 1999), and the Convention on Biological Diversity was signed at the Earth Summit in 1992, linking biodiversity preservation with indigenous knowledge. What these and many other state and corporate policy shifts have in common is a focus on, and increased legitimization of, environmental knowledge claims produced outside the academy.

The concurrent explosion of interest in extramural science in academia is striking both for its exponential character (98.8 percent of articles on extramural science included in the Web of Knowledge have been published since 1980 [Lave 2011]) and for its fragmentation into almost entirely disconnected sets of literatures. Some scholars address two forms of extramural science together (Fischer 2000; Graddy 2011; Harding 2008; Fairhead and Leach 2003; Leach and Fairhead 2002; and much of the Critical Geographic Information Systems (GIS) literature on Web 2.0); however, no one analyzes the range of types relationally as deeply interconnected phenomena.

This is a startling oversight for several reasons. First, despite the varying locations, political economies, and histories of these different knowledge sources, there is a great deal of overlap among them. Most share a topical focus on the environment (particularly local knowledge, indigenous knowledge, and both forms of citizen science). They have similarly informal, low-budget conditions of production: no Big Science here. Perhaps most important, these different forms of extramural science are united by their shared relegation to Western science's foil in any number of highly loaded binaries including: universal/particular, dynamic/static, disinterested/embedded, oral/written, cerebral/embodied, and analytical/intuitive.

A second set of reasons it is unhelpful to study these disparate forms of extramural knowledge production separately is the commonalities in how they have been treated by others. As I argue in more detail below, extramural knowledge producers, regardless of type, are a current focus of primitive accumulation. They have also been the focus of a dramatic resurgence of interest by policymakers and by academics. Natural scientists faced with increasing competition for grants as a result of reductions in public funding were in need of extensive, inexpensive assistance collecting data. For their part, critical social scientists were inspired to seek alternative sources of knowledge by postcolonial agendas, and by the search for counterclaims to block the neoliberal intensification of state and corporate appropriation of local natural resources in both developed and developing countries.

Extramural knowledge producers share a predominantly environmental focus, lack of institutional base, and illegitimacy compared to university-produced knowledge claims. They also have a shared status as a target of appropriation, and of academic study catalyzed by a variety of neoliberal forces. Thus despite their obvious differences, the commonalities among types of extramural science make it imperative to consider them together (Figure 4).

Historians of science and STS scholars, particularly those concerned with the tension between elitism and democracy (Reingold 1976), use the term *amateur scientist* to describe people who play a recognized role in scientific communities but are not full-time scientists. Amateur scientists are the most visible continuity from the integrated scientific community of the early 1800s, and are clustered in fields where requirements for entry are not so costly, such as astronomy

Figure 4. Facets of extramural knowledge production

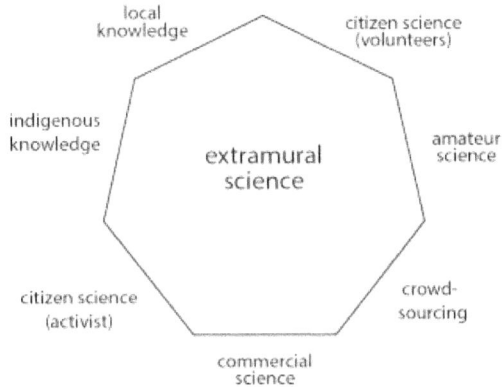

and archaeology. While the majority of amateur scientists hold relatively little legitimacy some, typically well-educated white-collar workers in developed countries, hold nearly professional status (Bhanoo 2011).

Anthropologists, geographers, and STS scholars use the term *indigenous knowledge* (also indigenous ecological knowledge, indigenous technical knowledge, or traditional ecological knowledge) to describe the agro-ecological knowledge of subsistence communities in geographically and politically marginal communities in the developing world (Dove 1996, 2006, and 2011; Harding 2008, 2011; Hayden 2003; Nadasdy 1999; Verran and Turnbull 1995). Indigenous knowledge producers are neither white nor white-collar, and are typically dismissed outside of the critical social sciences as holders of culturally embedded knowledge rather than active builders and curators of changing bodies of knowledge. Researchers of indigenous knowledge thus often take protective stances, analyzing these knowledge claims as targets of appropriation and agnotology (Brush and Stabinsky 1996; Greene 2004; Hobart 1993; Shiva 2001; Van der Ploeg 1993), critical sources of counterclaims (Bryan 2011; Davis 2007; Fairhead and Leach 1996; Rundstrom 1995), and beacons for sustainable practices (Mearns and Norton 2010; Nelson 2008; Reed 2009; Shaw et al. 2009).

STS scholars, sociologists, and environmental managers use the term *local knowledge* to describe similar agro-ecological knowledge when developed and held by geographically, economically, and politically marginal white people in developed countries (Fischer 2000; Irwin et al. 1996; Wynne 1996),[6] such as sheep farmers in Britain or heritage seed savers in Appalachia. As with indigenous knowledge systems, local knowledge systems are typically analyzed in the scholarly literature as marginalized, disregarded, or endangered (Fischer 2000; Irwin and Wynne 1996) or as critical resources for sustainable living (Graddy 2011; Irwin 1995).[7]

Crowdsourcing describes the emerging practice of distributed communal problem solving. Like amateur scientists, crowdsourcers are typically highly educated individuals without academic or corporate research jobs—"the kind of scientific talent and expertise that corporate America … [was not previously able] to tap" (Howe 2006). Corporations use crowdsourcing to address problems in-house R&D divisions have been unable to solve, most famously improving Netflix's recommendation algorithm (Thompson 2008). Though crowdsourcing has been addressed primarily in business, computing, and management journals (e.g., Elkins and Williams 2010; Leimeister et al. 2009; Terwiesch and Xu 2010), it is drawing increasing amounts of attention from critical GIS scholars intrigued by the democratizing potential of Web 2.0 cartography (Crampton 2009; Elwood 2008b; Goodchild 2007; Sui 2008; Wilson et al. 2009; Zook et al. 2010).

Citizen science has two quite different meanings. For natural scientists, citizen science describes the practice of enlisting large numbers of unpaid volunteers (typically white and well-educated) to collect data, most often for large-scale ecological or astronomical studies (Cohn 2008; Dickinson et al. 2010; Ellis and Waterton 2004; Greenwood 2005). By contrast, STS scholars use citizen science to denote practices of activist or counterscience centered in low-income communities, often of color, such as popular epidemiology, participatory mapping, and the knowledge production practices of the environmental justice movement more broadly (B. Allen 2003; Brown and Mikkelsen 1990; Corburn 2005; Craig et al. 2002; Elwood 2008a; Ottinger 2010).

Last, and closest to academic science, STS scholars, historians of science, anthropologists, and geographers study something that could be called *commercial science:* private sector knowledge claims developed in settings as disparate as Bell Labs, scrappy biotech start-ups, contract-research organizations, and even individual consultancies (Fisher 2009; Lave 2012a, 2012b; Mirowski and Van Horn 2005; Randalls 2010; Shapin 2008; Sunder-Rajan 2006). In the vast majority of these settings, knowledge claims are developed in formal settings by researchers with advanced academic degrees and high levels of scientific legitimacy. Like academic scientists, commercial scientists receive federal research funding, publish in academic journals, and present their findings at academic conferences (though with more severe intellectual property and secrecy constraints). These diverse forms of extramural knowledge have different geopolitical-economic positions, but a great deal of overlap in terms of the status of participants and the types of knowledge claims they typically produce. As I will argue in the next section, they have also experienced remarkably similar impacts from neoliberal science regimes.

Impacts of Neoliberal Science Regimes Outside Universities

Unsurprisingly, many scholars of commercial science are forcibly confronted with neoliberal emphases on the commercialization and privatization of knowledge production (Fisher 2009; Johnson 2009; Mirowski and Van Horn 2005; Randalls 2010). Similarly, many scholars of indigenous knowledge, where neoliberal strategies of primitive accumulation via biopiracy have had some impact, are also quite cognizant of neoliberal influences on their field sites (Bryan 2011; Greene 2004; Hayden 2003). By contrast, in the literatures on amateur science, local knowledge, crowdsourcing, and both types of citizen science there is as yet little discussion of neoliberalism, which seems to be viewed as irrelevant to the micropolitics and struggles for legitimacy of knowledge developed outside the academy. But while neoliberal science regimes' influences on knowledge produced outside the university are different from those inside the academy, they are just as striking.

A central impact is a new wave of appropriation of labor and knowledge. Both amateur and citizen scientists provide vast amounts of unpaid work for physical scientists. Citizen science projects, in particular, are increasingly common; the Cornell Laboratory of Ornithology (CLO), the citizen science mothership in the United States, lists more than 600 volunteer-staffed projects (Dickinson et al. 2010: 150). As CLO researchers wrote in a recent review article, "The bang for the buck can be good" (ibid.: 151). Their Project FeederWatch, for example, is supported by volunteers who not only pay fees to participate, but also contribute an estimated $3 million per year in unpaid labor. Environmental scientists are even reaching back into historical records to posthumously convert amateur naturalists into citizen scientists, most visibly in the case of Thoreau (Primack et al. 2009). And while crowdsourcers may be paid, they are employed only on an as-needed basis and have no job security or benefits. The winners of big prizes such as the Netflix challenge get the most press, but it is important to note that the vast majority of crowdsourcing contracts are far less remunerative (Howe 2006).

In addition to labor, there is also a new wave of appropriation of both indigenous and local knowledge, typically carried out by ethnobotanists and other bioprospectors in the service of pharmaceutical and agricultural biotechnology companies. Arturo Escobar described this process of appropriation in a classic 1996 article:

> This second form of capital relies not only on the symbolic conquest of nature (in terms of "biodiversity reserves") and local communities (as "stewards" of nature); it also requires the semiotic conquest of local knowledges, to the extent that "saving nature" demands the valuation of local knowledges of sustaining nature. ... This triple cultural reconversion of nature, people, and knowledge represents a novel internalization of production conditions. Nature and local people themselves are seen as the source and creators of value. (Escobar 1996: 57)

In striking contrast to these destructive dynamics of appropriation, neoliberal science regimes' privileging of privatized and commercializable knowledge has also generated a big boost in the credibility of extramural science and the people who produce it, with contradictory effects. On the one hand, commercial knowledge claims are no longer seen as potentially compromised by conflicts of interest, but instead as legitimized by market forces. On the other hand, environmental consultants have been able to establish serious scientific credibility in startling upsets of the traditional construction of scientific legitimacy (Briske et al. 2011; Lave 2012a, 2012b). Citizen scientists (of the activist persuasion) are now accorded a place at the table in many regulatory decisions (Brown and Mikkelsen 1990; Cohen and Ottinger 2011; Frickel et al. 2010; Moore et al. 2011; Ottinger 2009), and indigenous groups are consulted in developing environmental management plans (Fairhead and Leach 2003; Nadasdy 1999) in ways that were unimaginable in the 1950s, or even in the 1970s.

That neoliberal science regimes' impacts—positive and negative—are so unified across these seemingly disparate forms of extramural knowledge production, and yet so little remarked on in the scholarly literature strongly suggests that the academic boundaries among research on types of nonacademic knowledge production are not just artificial, but actively unhelpful. When looked at comparatively, trends invisible to researchers of particular extramural knowledge forms become strikingly clear.

Larger Implications

I have argued that political-economic forces shape not just the application of environmental knowledge claims, but also their production (see Figure 3), as conditions of practice both inside and outside the academy are increasingly influenced by neoliberal science regimes and knowledge producers' responses to them. Outside the academy, neoliberal science regimes have so far had interestingly contradictory affects: appropriation, privatization, and commercialization, on the one hand; increased visibility and respect, on the other. Do more credibility and a place at the grown-ups' table translate directly into autonomy, legitimacy, and vastly improved conditions for marginalized communities in the developing and developed worlds? Clearly not. But the hugely increased level of attention from policymakers, academics, and corporations is a notable shift from centuries of disregard and disempowerment, and may enable extramural knowledge producers sufficient legitimacy to participate in the production of credible scientific claims.

Inside the academy, however, the reported impacts of neoliberal science regimes have thus far been quite grim. Graduates are marooned with heavy debt loads at the same time that the devaluation of teaching seems to be reducing the quality of the education they receive. Peer review has long been flawed by cronyism, laziness, and elitism, but is delegating intellectual gatekeeping to corporate interests really preferable? The evisceration of funding for the humanities, much of the social sciences, and basic science research is deeply distressing, and it is increasingly clear that the dramatic expansion of intellectual property protections is reducing research productivity rather than increasing it (e.g., Henry et al. 2003; Mirowski 2011: 140, fn1; Pestre 2005: 35; Rodriguez et al. 2007).

Impacts from neoliberal science regimes thus matter deeply to knowledge producers inside and outside universities worldwide as they radically and abruptly reshape our conditions of practice. It is important to note that the impacts do not stop there: the transformation of the organization, practice, and content of science has larger social justice impacts because of how such knowledge is circulated (or not) and how it is applied.

One issue arises from limitations on access to privatized knowledge and to research tools sequestered behind intellectual property firewalls. For example, Leigh Johnson has described how climate researchers at an American university created a spin-off business that developed hurricane forecasts for a major energy company. These privately held forecasts can predict hurricane paths seven days in advance within a range of 100 miles, while the publicly available forecasts developed by the US National Hurricane Center achieve the same track accuracy only 48 hours in advance (Johnson 2009). Imagine what might have been accomplished with public access to accurate forecasts of the paths of Hurricanes Katrina and Rita seven days in advance; at present, only a major energy company has access to such knowledge. Jill Fisher (2009) has demonstrated that clinical trials in the United States are largely filled with low-income patients who not only will never be able to afford the patented medication if the trial is successful, but who also would not otherwise have access to health care of any kind. The same players that shut low-income Americans out of access to health care—the insurance and pharmaceutical industries—generate profit by using their bodies to produce scientific claims. Kaushik Sunder-Rajan (2006) describes similar impacts from the privatized production of clinical trial data in India, with the added kick that the construction of medical centers and clinical trial facilities expropriates the land of people whose best employment prospect then becomes "volunteering" to be paid clinical trial participants.

A second issue arises from the ways in which privatized and/or commercialized knowledge can enable the neoliberalization of other realms. For example, the basic principle of neoliberal environmental management is to establish markets that offset environmental harms by internalizing their cost. But getting such markets up and running requires converting complex ecosystems services into tradable commodities. Without metrics that pare away the ecological specificity of particular ecosystems in order to abstract them into easily measurable, comparable units, market-based environmental management cannot function (Lave et al. 2010a; Robertson 2006). Creating these market-enabling metrics is thus a critical task of "science in the service of capital" (Robertson 2006).

In stream mitigation banking (SMB), an increasingly common form of market-based environmental management, for-profit bankers speculatively restore damaged streams to create a bank of "credits" that developers can buy in order to obtain a permit that allows them to destroy an inconveniently located stream elsewhere. SMB was established and spread very rapidly because there was a commodity-defining metric ready to hand: a privately produced stream classification system that was itself a product of neoliberal promotion of the privatization and commercialization of science (Lave 2012a, 2012b; Lave et al. 2010a). By market standards, this metric proved quite practical, enabling the stream mitigation market to function smoothly. By ecological standards, however, SMB has thus far failed because streams restored to create mitigation credits show no substantial improvement in ecological function or water quality (Bernhardt and Palmer 2011; Doyle and Shields 2012). The privately produced, market-ready knowledge claims created as science was filtered through the sieve of commercial purposes—to paraphrase Gibbons et al. (2004: 163)—have enabled the spread of market-based environmental management and a new driver of environmental damage to streams. The broader Payment for Ecosystem Services paradigm of which mitigation banking is a part seems similarly vulnerable to the synergistic intersection of neoliberal science and environmental management regimes.

Obviously, the intersection of environmental damage and injustice is not new; the historians among us can provide all manner of heart-wrenching examples of environmental harms justified or hidden by past political-economic orders, from Bikini Island to Bhopal. What is novel is the market-based form these impacts take today, which raise particular kinds of challenges for critical nature/society scholars.

What then shall we do? These specific examples and the discussion that preceded them have clear intellectual implications for critical nature/society studies. In order to understand the dynamic application of neoliberal environmental policies and management practices, we must expand our research to include the interlinked production and circulation of the science that enables them. I would argue that we should also respond to neoliberal science regimes as activists: the neoliberalization of knowledge production has social justice implications that extend far beyond our livelihoods.

The first order of business is to decide what exactly we want to defend. As many have pointed out, before the advent of neoliberal science regimes the university was hardly Edenic (Apple 2005; Mirowski 2011; Slaughter and Rhoades 2004). Slaughter and Rhoades write that: "The not-too-distant past in higher education (like the continued present) featured fundamental social inequities, significant constraints on the free pursuit of knowledge, [and] a linking of the research enterprise to the purposes and mechanisms of the cold war" (2004: 33). Thus as Michael Apple wisely notes, we must be very careful about what parts of the old systems of higher education we choose to advocate (2005: 24).

Beyond this critical task of viewing the past without nostalgia, Dominique Pestre argues that we need to carefully evaluate neoliberal science regimes because we could actually be ambivalent about them if we chose to:

> First because operationality and pragmatism are not uninteresting criteria, far from it, criteria that we could just ignore or dismiss lightly; but also, and more profoundly, because the universal, epistemological and moral, values of science as "pure" knowledge and culture— notions and values that were invented in several steps from the late XVIIth to the late XIXth centuries to differentiate "us-in-the-West" from "them-in-the-Orient," and "us-the-scientists" from "them-the-laypeople"— are not indisputable and without drawbacks. Heavily loaded, "ideologically speaking," we cannot just accept them as non-problematic. (2005: 38)

Supporting this point, it is worth remembering that neoliberalization is not simply a result of top-down, structured processes, but also of individuals finding worth in and embracing particular components of neoliberal philosophies (Larner 2003). Some extramural knowledge producers have used the drive to commercialize their knowledge claims as a source of political leverage (as demonstrated by Tania Li's [2000] powerful analysis of indigenous groups choosing the "tribal slot"), and as Slaughter and Rhoades (2004) and Nowotny (2005) note, many academic administrators and researchers (including some with progressive agendas) have embraced greater connection to markets and fought for policy changes to enable them. If we took these arguments for ambivalence toward neoliberal science regimes seriously, our political agenda would include not just targeted rejection of particular aspects of neoliberal science management, but also re-appropriation of some of its core practices, such as choice, accountability, and "relevance," to more progressive ends.

Neoliberal science regimes have direct social and environmental impacts. As knowledge claims produced under neoliberalized conditions are circulated and applied they advance commercial interests, heighten the impacts of social inequality, and enable the neoliberalization of as yet un(der)capitalized realms. The same forces we see at work in our field sites shape the science underlying the policies we critique, as well as the counter-science we sometimes use to oppose them. We can no longer afford to ignore their interconnections in either our intellectual or political practice.

◾ ACKNOWLEDGMENTS

I am grateful to Joe Bryan, Gwen Ottinger, Morgan Robertson, and Matt Wilson for bibliographic tour guiding in their areas of expertise, and to Ilana Gershon for bringing the CFP for this issue to my attention. The argument I present here has been much strengthened by conversations with Phil Mirowski, Jason Moore, Gwen Ottinger, Sam Randalls, and Matt Wilson, and by feedback from audiences at the Association of American Geographers annual meeting; the Sawyer Seminar "Rupture and Flow: The Circulation of Technoscientific Objects and Facts" at Indiana University; the Studies in Science, Technology, and Environment program at Umeå University in Sweden; the Nature, Inc. conference in The Hague; the Geography Department at the University of Kentucky; and the Society for the Social Studies of Science annual meeting, all in 2010. Finally, Jean Lave and Matt Wilson provided very helpful critiques at the rough draft stage.

REBECCA LAVE is an assistant professor in the Geography Department at Indiana University. Her research draws together political ecology, STS, and fluvial geomorphology as part of the emerging subfield Critical Physical Geography. Her recent work focuses on the political economy of environmental knowledge production, and includes a book (*Fields and Streams: Stream Restoration, Neoliberalism and the Future of Environmental Science* [2012]) and articles in the *Annals of the Association of American Geographers*, *Ecological Restoration*, the *Journal of the American Water Resources Association*, and *Social Studies of Science*.

◾ NOTES

1. http://www.cbpp.org/cms/index.cfm?fa=view&id=1214 (accessed 14 November 2011).
2. http://www.prweb.com/releases/DynaMed/F1000/prweb8948133.htm (accessed 13 December 2011).
3. As Mirowski has noted (2011), patents are only one component of this broad expansion of intellectual property rights which also includes material transfer agreements (MTAs) and the extension of copyright and trade secrets provisions. MTAs, in particular, have become the intellectual property protection tool of choice as they are faster, easier, and cheaper to establish than patents.
4. I have borrowed the term *extramural science* from Ronald Barnett (2005). While I do not find it particularly evocative, I have yet to discover a better umbrella term for the wide range of knowledges developed outside the academy.
5. And in some cases racial and gender differences as well. The history of indigenous knowledge under colonialism often featured far sharper status distinctions than the predominantly Western history I describe here (e.g., Michael Adas' classic work *Machines as the Measure of Men*). Michael Dove (2011) has noted a more benign separation between indigenous and colonial knowledge producers, as mutual ignorance and imagination facilitated trade relations.

6. The strong constructivist positions of many STS scholars leads them to describe all science as local knowledge (Verran and Turnbull 1996; Wynne 1996: 382). Also, some scholars uncomfortable with privileging autochthony and the severe limitations that can accompany claims to indigeneity use *local knowledge* to describe knowledge production among groups typically considered indigenous (Evans 2011; Mutersbaugh 2006).

7. The age of the STS cites is telling: while local knowledge was prominent in the 1990s, it is not even included in the index of the current 2007 edition of the *Handbook of Science and Technology Studies*. The attention previously devoted to local knowledge seems to have gone into activist science and debates about the democratization of expertise.

■ **REFERENCES**

Adas, Michael. 1989. *Machines as the Measure of Men: Science, Technology, and Ideologies of Western Dominance.* Ithaca, NY: Cornell University Press.

Allen, Barbara. 2003. *Uneasy Alchemy: Citizens and Experts in Louisiana's Chemical Corridor Disputes.* Cambridge: MIT Press.

Allen, David Elliston. 1976. *The Naturalist in Britain.* London: Penguin.

Apple, Michael W. 2005. "Education, Markets, and an Audit Culture." *Critical Quarterly* 47, nos. 1–2: 11–29.

Barnett, Ronald. 2005. "Re-opening Research: New Amateurs or New Professionals?" Pp. 263–277 in *Participating in the Knowledge Society,* ed. Ruth Finnegan. New York: Palgrave Macmillan.

Bernhardt, Emily, and Margaret A. Palmer. 2011. "River Restoration: the fuzzy logic of repairing reaches to reverse catchment scale degradation." *Ecological Applications* 21, no. 6: 1926–1931.

Bhanoo, Sindya N. 2011. "Dinosaur-Hunting Hobbyist Makes Fresh Tracks for Paleontology." *New York Times,* 28 February.

Biagioli, Mario. 2006. "Patent Republic: Specifying Inventions, Constructing Authors and Rights." *Social Research* 73, no. 4: 1129–1172.

Blaikie, Piers. 1985. *The Political Economy of Soil Erosion in Developing Countries.* New York: John Wiley & Sons.

Bourdieu, Pierre. 1998. "Rethinking the State: Genesis and Structure of the Bureaucratic Field." Pp. 35–63 in *Practical Reason: On the Theory of Action,* ed. Pierre Bourdieu. Stanford, CA: Stanford University Press.

Braun, Bruce. 2002. *The Intemperate Rainforest: Nature, Culture and Power on Canada's West Coast.* Minneapolis: University of Minnesota Press.

Briske, D.D., Nathan F. Sayre, Lynn Huntsinger, M. Fernandez-Gimenez, B. Budd, and J.D. Derner. 2011. "Origin, Persistence, and Resolution of the Rotational Grazing Debate: Integrating Human Dimensions into Rangeland Research." *Rangeland Ecology & Management* 64, no. 4: 325–334.

Brown, Phil, and Edwin Mikkelsen. 1990. *No Safe Place: Toxic Waste, Leukemia, and Community Action.* Berkeley: University of California Press.

Brush, Stephen B., and Doreen Stabinsky. 1996. *Valuing Local Knowledge: Indigenous People and Intellectual Property Rights.* Washington, DC: Island Press.

Bryan, Joseph. 2011. "Walking the Line: Participatory Mapping, Indigenous rights, and Neoliberalism." *Geoforum* 42, no. 1: 40–50.

Canaan, Joyce E., and Wesley Shumar. 2008. "Higher Education in the Era of Globalization and Neoliberalism." Pp. 1–30 in *Structure and Agency in the Neoliberal University,* ed. Joyce E. Canaan and Wesley Shumar. London: Routledge.

Cohen, Benjamin, and Gwen Ottinger. 2011. *Technoscience and Environmental Justice.* Cambridge: MIT Press.

Cohn, Jeffrey P. 2008. "Citizen Science: Can Volunteers Do Real Research?" *Bioscience* 58, no. 3: 192–197.

Corburn, Jason. 2005. *Street Science: Community Knowledge and Environmental Health Justice.* Cambridge: MIT Press.

Coriat, Benjamin, and Fabienne Orsi. 2002. "Establishing a New Intellectual Property Rights Regime in the United States." *Research Policy* 31: 1491–1507.

Craig, William J., Trevor M. Harris, and Daniel Weiner. 2002. *Community Participation and Geographical Information Systems.* New York: Taylor & Francis.

Crampton, Jeremy W. 2009. "Cartography: Maps 2.0." *Progress in Human Geography* 33, no. 1: 91–100.

Davis, Diana K. 2007. *Resurrecting the Granary of Rome: Environmental History and French Colonial Expansion in North Africa.* Athens: Ohio University Press.

Demeritt, David. 1998. "Science, Social Constructivism and Nature." Pp.173–193 in *Remaking Reality: Nature at the Millennium,* ed. Bruce Braun and Noel Castree. New York: Routledge.

Demeritt, David. 2001. "The Construction of Global Warming and the Politics of Science." *Annals of the Association of American Geographers* 91, no. 2: 307–337.

Demeritt, David. 2006. "Science Studies, Climate Change, and the Prospects for Constructivist Critique." *Economy and Society* 35, no. 3: 453–479.

Dickinson, Janis L., Benjamin Zuckerberg, and David N. Bonter. 2010. "Citizen Science as an Ecological Research Tool: Challenges and Benefits." *Annual Review of Ecology, Evolution, and Systematics* 41: 149–172.

Dove, Michael. 1996. "Center, Periphery, and Biodiversity: A Paradox of Governance and a Developmental Challenge." Pp. 41–67 in *Valuing Local Knowledge,* ed. Stephen B. Brush and Doreen Stabinsky. Washington, DC: Island Press.

Dove, Michael. 2006. "Indigenous Peoples and Environmental Politics." *Annual Review of Anthropology* 35: 191–208.

Dove, Michael. 2011. *The Banana Tree at the Gate: A History of Marginal Peoples and Global Markets in Borneo.* New Haven, CT: Yale University Press.

Doyle, Martin W., and F. Douglas Shields. 2012. "Compensatory Mitigation for Streams Under the Clean Water Act: Reassessing Science and Redirecting Policy`." *Journal of the American Water Resources Association* 48, no. 3: 494–509.

Duvall, Chris. 2011. "Ferricrete, Forests, and Temporal Scale in the Production of Colonial Science in Africa." Pp. 113–127 in *Knowing Nature: Conversations at the Border of Political Ecology and Science Studies,* ed. Mara Goldman, Paul Nadasdy, and Matt Turner. Chicago: University of Chicago Press.

Elkins, Sean, and Antony J. Williams. 2010. "Reaching Out to Collaborators: Crowdsourcing for Pharmaceutical Research." *Pharmaceutical Research* 27, no. 3: 393–395.

Ellis, Rebecca, and Claire Waterton. 2004. "Environmental Citizenship in the Making: The Participation of Volunteer Naturalists in UK Biological Recording and Biodiversity Policy." *Science and Public Policy* 31, no. 2: 95–105.

Elwood, Sarah. 2008a. "Volunteered Geographic Information: Future Research Directions Motiviated by Critical, Participatory, and Feminist GIS." *GeoJournal* 72, nos. 3–4: 173–183.

Elwood, Sarah. 2008b. "Volunteered Geographic Information: Key Questions, Concepts and Methods to Guide Emerging Research and Practice." *GeoJournal* 72, nos. 3–4: 133–135.

Escobar, Arturo. 1996. "Constructing Nature: Elements for a Poststructural Political Ecology." Pp. 46–68 in *Liberation Ecologies: Environment, Development, Social Movements,* ed. Richard Peet and Michael Watts. London: Routledge.

Evans, Laura. 2011. *Power from Powerlessness: Tribal Governments, Institutional Niches, and American Federalism.* Oxford: Oxford University Press.

Fairhead, James, and Melissa Leach. 1996. "Rethinking the Forest-Savanna Mosaic: Colonial Science & its Relics in West Africa." Pp. 105–121 in *The Lie of the Land: Challenging Received Wisdom on the African Environment,* ed. Melissa Leach and Robin Mearns. Oxford: James Currey.

Fairhead, James, and Melissa Leach. 2003. *Science, Society and Power: Environmental Knowledge and Policy in West Africa and the Caribbean.* Cambridge: Cambridge University Press.

Fischer, Frank. 2000. *Citizens, Experts, and the Environment: The Politics of Local Knowledge.* Durham, NC: Duke University Press.

Fisher, Jill A. 2009. *Medical Research for Hire: the Political Economy of Pharmaceutical Clinical Trials.* New Brunswick, NJ: Rutgers University Press.

Forsyth, Tim. 2003. *Critical Political Ecology: The Politics of Environmental Science.* London: Routledge.

Frickel, Scott, Sahra Gibbon, Jeff Howard, Joana Kempner, Gwen Ottinger, and David J. Hess. 2010. "Undone Science: Social Movement Challenges to Dominant Scientific Practice." *Science, Technology & Human Values* 35, no. 4: 444–473.

Geiger, Roger, and Creso Sa. 2008. *Tapping the Riches of Science: Universities and the Promise of Economic Growth.* Cambridge, MA: Harvard University Press.

Gibbons, Michael, Camille Limoges, Helga Nowotny, Simon Schwartzman, Peter Scott, and Martin Trow. 1994. *The New Production of Knowledge: The Dynamics of Science and Research in Contemporary Societies.* London: Sage.

Goldman, Mara, and Matthew D. Turner. 2011. "Introduction." Pp. 1–23 in *Knowing Nature: Conversations at the Intersection of Political Ecology and Science Studies,* ed. Mara Goldman, Paul Nadasdy and Matthew D. Turner. Chicago: University of Chicago Press.

Goodchild, Michael F. 2007. "Citizens as Sensors: the World of Volunteered Geography." *GeoJournal* 69, no. 4: 211–221.

Graddy, T. Garrett. 2011. Situating *in situ:* Agrobiodiversity Conservation & Regeneration in Appalachian Kentucky and the Peruvian Andes. PhD diss., University of Kentucky, Lexington.

Grant, Bob. 2009. "Elsevier Published 6 Fake Journals." *Scientist,* 7 May.

Gravois, John. 2006. "Tracking the Invisible Faculty." *Chronicle of Higher Education* 53, no. 17: A8–A9.

Greene, L. Shane. 2004. "Indigenous People Incorporated? Culture as Politics, Culture as Property in Bioprospecting." *Current Anthropology* 45, no. 2: 211–237.

Greenwood, Jeremy J.D. 2005. "Science with a Team of Thousands: The British Trust for Ornithology." Pp. 152–165 in *Participating in the Knowledge Society,* ed. Ruth Finnegan. New York: Palgrave Macmillan.

Guthman, Julie, and Melanie Dupuis. 2006. "Embodying Neoliberalism: Economy, Culture, and the Politics of Fat." *Environment and Planning D: Society and Space* 24, no. 3: 427–448.

Hahn, Ryan. 2007. The Global State of Higher Education and the Rise of Private Finance. Washington, D.C.: Institute for Higher Education Policy.

Harding, Sandra. 2008. *Science from Below: Feminisms, Postcolonialities, and Modernities.* Durham, NC: Duke University Press.

Harding, Sandra. 2011. *The Postcolonial Science and Technology Studies Reader.* Durham, NC: Duke University Press.

Hayden, Cori. 2003. *When Nature Goes Public: The Making and Unmaking of Bioprospecting in Mexico.* Princeton, NJ: Princeton University Press.

Hecht, Susanna. 1985. "Environment, Development and Politics: Capital Accumulation and the Livestock Sector in Eastern Amazonia." *World Development* 13, no. 6: 663–684.

Henry, Michelle, Mildred Cho, Meredith Weaver, and Jon Merz. 2003. "A Pilot Survey of the Licensing of DNA Inventions." *Journal of Law, Medicine and Ethics* 31: 442.

Hobart, Mark. 1993. *An Anthropological Critique of Development: the Growth of Ignorance.* London: Routledge.

Hollander, Gail M. 2008. *Raising Cane in the 'Glades: The Global Sugar Traffic and the Transformation of Florida.* Chicago: University of Chicago Press.

Howe, Jeff. 2006. The Rise of Crowdsourcing. *Wired.*

Irwin, Alan. 1995. *Citizen Science: A Study of People, Expertise and Sustainable Development.* London: Routledge.

Irwin, Alan, and Brian Wynne. 1996. "Introduction." Pp. 1–18 in *Misunderstanding Science? The Public Reconstruction of Science and Technology,* ed. Alan Irwin and Brian Wynne. Cambridge: Cambridge University Press.

Irwin, Alan, Alison Dale, and Dennis Smith. 1996. "Science and Hell's Kitchen: the Local Understanding of Hazard Issues." Pp. 47–64 in *Misunderstanding Science? The Public Reconstruction of Science and Technology,* ed. Alan Irwin and Brian Wynne. Cambridge: Cambridge University Press.

Johnson, Leigh. 2009. "Scientific Diligence, Climatic Urgency, and the Public/Private Expert: Hurricane Forecasts Go to Market." Paper presented at the Society for the Social Study of Science conference, Washington DC.

Kaye, Tim, Robert Bickel, and Tim Birtwistle. 2006. "Criticizing the Image of the Student as Consumer: Examining Legal Trends and Administrative Responses in the US and UK." *Education & the Law* 18 no. 2–3: 85–129.

Keeney, Elizabeth B. 1992. *The Botanizers: Amateur Scientists in Nineteenth Century America.* Chapel Hill: University of North Carolina Press.

Kirsch, Stuart. 2011. "Science as Corporate Strategy." Paper presented at *History Underground: Environmental Perspectives on Mining,* organized by Frank Uekotter and John McNeill. Rachel Carson Center, University of Munich.

Kleinman, Daniel Lee. 2003. *Impure Cultures: University Biology and the World of Commerce.* Madison: University of Wisconsin Press.

Knell, Simon J. 2000. *The Culture of English Geology, 1815–1851: A Science Is Revealed Through Its Collecting.* Burlington, VT: Ashgate.

Kohler, Robert E. 2006. *All Creatures: Naturalists, Collectors, and Biodiversity, 1850–1950.* Princeton, NJ: Princeton University Press.

Lambert, Cath, Andrew Parker, and Michael Neary. 2007. "Entrepreneurialism and Critical Pedagogy: Reinventing the Higher Education Curriculum." *Teaching in Higher Education* 12, no. 4: 535–537.

Larner, Wendy. 2003. "Neoliberalism?" *Environment and Planning D: Society and Space* 21, no. 5: 509–512.

Lave, Rebecca. 2011. "Neoliberalism and Knowledge from the Margins." Paper presented at Society for Social Studies of Science, 2 November, Cleveland, OH.

Lave, Rebecca. 2012a. "Bridging Political Ecology and STS: A Field Analysis of the Rosgen Wars." *Annals of the Association of American Geographers* 102, no. 2: 366–382.

Lave, Rebecca. 2012b. *Fields and Streams: Stream Restoration, Neoliberalism, and the Future of Environmental Science.* Athens: University of Georgia Press.

Lave, Rebecca, Martin W. Doyle, and Morgan M. Robertson. 2010a. "Privatizing Stream Restoration in the U.S." *Social Studies of Science* 40, no. 5: 677–703.

Lave, Rebecca, Philip Mirowski, and Samuel Randalls. 2010b. "Introduction: STS and Neoliberal Science." *Social Studies of Science* 40, no. 5: 659–675.

Leach, Melissa, and James Fairhead. 2002. "Manners of Contestation: 'Citizen Science' and 'Indigenous Knowledge' in West Africa and the Caribbean." *International Social Science Journal* 54, no. 173: 299–312.

Leach, Melissa, and Robin Mearns. 1996. *The Lie of the Land: Challenging Received Wisdom on the African Environment.* Portsmouth, NH: Heinemann.

Leimeister, Jan Marco, Michael Huber, and Ulrich Bretschneider. 2009. "Leveraging Crowdsourcing: Activation-Supporting Components for IT-Based Ideas Competition." *Journal of Management Information Systems* 26, no. 1: 1997–1224.

Li, Tania Murray. 2000. "Articulating Indigenous Identity in Indonesia: Resource Politics and the Tribal Slot." *Comparative Studies in Society and History* 42, no. 1: 149–179.

Macilwain, Colin. 2011. "'Facebook of Science' Seeks to Reshape Peer Review." *Chronicle of Higher Education.* 57, no. 22.

Martin, Andrew, and Andrew W. Lehren. 2012. "A Generation Hobbled by the Soaring Cost of College." *New York Times,* 12 May.

McCarthy, James. 2002. "First World Political Ecology: Lessons from the Wise Use Movement." *Environmental and Planning A* 34, no. 7: 1281–1302.

Mearns, Robin, and Andrew Norton. 2010. *Social Dimensions of Climate Change: Equity and Vulnerability in a Warming World.* Washington, DC: World Bank.

Mirowski, Philip. 2011. *Science-Mart: Privatizing American Science.* Cambridge, MA: Harvard University Press.

Mirowski, Philip, and Robert Van Horn. 2005. "The Contract Research Organization and the Commercialization of Scientific Research." *Social Studies of Science* 35, no. 4: 503–548.

Moore, Kelly, Daniel Lee Kleinman, David J. Hess, and Scott Frickel. 2011. "Science and Neoliberal Globalization: A Political Sociological Approach." *Theory and Society* 40, no. 5: 505–532.

Mutersbaugh, Tad. 2006. "Certifying Biodiversity: Conservation Networks, Landscape Connectivity, and Certified Agriculture in Southern Mexico." Pp. 49–70 in *Globalization & New Geographies of Conservation,* ed. Karl Zimmerer. Chicago: University of Chicago Press.

Nadasdy, Paul. 1999. "The Politics of TEK: Power and the 'Integration' of Knowledge." *Arctic Anthropology* 36, nos. 1–2: 1–18.

Nedeva, Maria, and Rebecca Boden. 2006. "Changing Science: The Advent of Neoliberalism." *Prometheus* 24, no. 3: 269–281.

Nelson, Melissa K. 2008. *Original Instructions: Indigenous Teachings for a Sustainable Future.* Rochester, VT: Bear & Company.

Neumann, Roderick P. 1998. *Imposing Wilderness: Struggles over Livelihood and Nature Preservation in Africa.* Berkeley: University of California Press.

Nowotny, Helga. 2005. "The Changing Nature of Public Science." Pp. 1–28 in *The Public Nature of Science Under Assault,* ed. Helga Nowotny, Dominique Pestre, Eberhard Schmidt-Assman, Helmuth Schultze-Fielitz and Hans-Heinrich Trute. Berlin: Springer.

Nowotny, Helga, Peter Scott, and Michael Gibbons. 2001. *Rethinking Science: Knowledge and the Public in an Age of Uncertainty.* Cambridge: Polity Press.

Oleson, Alexandra, and Sanborn C. Brown. 1976. *The Pursuit of Knowledge in the Early American Republic: American Scientific and Learned Societies from Colonial Times to the Civil War.* Baltimore, MD: Johns Hopkins University Press.

Ottinger, Gwen. 2009. "Epistemic Fencelines: Air Monitoring Instruments and Expert-Resident Boundaries." *Spontaneous Generations* 3, no. 1: 55–67.

Ottinger, Gwen. 2010. "Buckets of Resistance: Standards and the Effectiveness of Citizen Science." *Science, Technology & Human Values* 35, no. 2: 244–270.

Pestre, Dominique. 2003. "Regimes of Knowledge Production in Society: Towards a More Political and Social Reading." *Minerva* 41, no. 3: 245–261.

Pestre, Dominique. 2005. "The Technosciences Between Markets, Social Worries and the Political: How to Imagine a Better Future?" Pp. 29–52 in *The Public Nature of Science Under Assault: Politics, Markets, Science and the Law,* ed. Helga Nowotny, Dominique Pestre, Eberhard Schmidt-Assman, Helmuth Schultze-Fielitz, and Hans-Heinrich Trute. Berlin: Springer.

Popp-Berman, Elizabeth. 2008. "Why Did Universities Start Patenting? Institution Building and the Road to the Bayh-Dole Act." *Social Studies of Science* 38, no. 6: 835–872.

Primack, Richard B., A Miller-Rushing, and K Dharaneeswaran. 2009. "Changes in the Flora of Thoreau's Concord." *Biological Conservation* 142 (3): 500–508.

Prudham, W. Scott. 2005. *Knock on Wood: Nature as Commodity in Douglas-Fir Country.* New York: Routledge.

Raffles, Hugh. 2002. *In Amazonia: A Natural History.* Princeton, NJ: Princeton University Press.

Randalls, Samuel. 2010. "Weather Profits: Weather Derivatives and the Commercialization of Meteorology." *Social Studies of Science* 40, no. 5: 705–730.

Reed, Richard K. 2009. *Forest Dwellers, Forest Protectors: Indigenous Models for International Development.* Upper Saddle River, NJ: Pearson Prentice Hall.

Reingold, Nathan. 1976. "Definitions and Speculations: The Professionalization of Science in America in the Nineteenth Century." Pp. 33–70 in *The Pursuit of Knowledge in the Early American Republic,* ed. Alexandra Oleson and Sanborn C. Brown. Baltimore, MD: Johns Hopkins University Press.

Rizzo, Michael. 2004. *A (Less Than) Zero Sum Game? State Funding for Public Education: How Public Higher Education Institutions Have Lost.* Dissertation, Cornell University, Ithaca, NY.

Robertson, Morgan M. 2006. "The Nature That Capital Can See: Science, State and Market in the Commodification of Ecosystem Services." *Environment and Planning D: Society and Space* 24, no. 3: 367–387.

Rodriguez, V., F. Janssens, K. Debackere, and B. DeMoor. 2007. "Do Material Transfer Agreements Affect the Choice of Research Projects?" *Scientometrics* 71, no. 2: 239–269.

Rundstrom, Robert A. 1995. "GIS, Indigenous Peoples, and Epistemological Diversity." *Cartography and Geographic Information Systems* 22, no. 1: 45–57.

Sayre, Nathan F. 2002. *Ranching, Endangered Species, and Urbanization in the Southwest: Species of Capital.* Tucson: University of Arizona Press.

Secord, Anne. 1994. "Science in the Pub: Artisan Botanists in Early 19th Century Lancashire." *History of Science* 32, no. 3: 269–315.

Shapin, Steven. 2008. *The Scientific Life: A Moral History of a Late Modern Vocation.* Chicago: University of Chicago Press.

Shaw, Rajib, Anshu Sharma, and Yukiko Takeuchi. 2009. *Indigenous Knowledge and Disaster Risk Reduction: From Practice to Policy.* New York: Nova Science.

Shiva, Vandana. 2001. *Protect or Plunder? Understanding Intellectual Property Rights.* London: Zed Books.

Singer, Natasha. 2009. "Merck Paid for Medical 'Journal' Without Disclosure." *New York Times,* 14 May.

Sismondo, Sergio. 2007. "From Ghost Writing to Ghost Management: How Much of the Medical Literature is Shaped behind the Scenes by the Pharmaceutical Industry?" *PLoS Medicine* 4, no. 9: e286.

Sismondo, Sergio. 2009. "Ghosts in the Machine: Publication Planning in the Medical Sciences." *Social Studies of Science* 39, no. 6: 171–198.

Slaughter, Sheila, and Gary Rhoades. 2004. *Academic Capitalism and the New Economy.* Baltimore, MD: Johns Hopkins University Press.

Strathern, Marilyn. 2000. *Audit Cultures: Anthropological Studies in Accountability, Ethics, and the Academy.* London: Routledge.

Sui, Daniel Z. 2008. "The Wikification of GIS and its Consequences: Or Angelina Jolie's New Tatoo and the Future of GIS." *Computers, Environment and Urban Systems* 32, no. 1: 1–5.

Sunder-Rajan, Kaushik. 2006. *Biocapital: The Constitution of Postgenomic Life.* Durham, NC: Duke University Press.

Terwiesch, Christian, and Yi Xu. 2008. "Innovation Contests, Open Innovation, and Multiagent Problem Solving." *Management Science* 54, no. 9: 1529–1543.

Thompson, Clive. 2008. "If You Liked This, You're Sure to Love That." *New York Times,* 21 November.

Turner, Matthew. 1999. "Conflict, Environmental Change and Social Institutions in Dryland Africa." *Society and Natural Resources* 12, no. 7: 643–657.

Tyfield, David. 2010. "Neoliberalism, Intellectual Property and the Global Knowledge Economy." Pp. 60–76 in *The Rise and Fall of Neoliberalism: The Collapse of an Economic Order?* ed. Kean Birch and Vlad Mykhenko. London: Zed Books.

Van der Ploeg, Jan Douwe. 1993. "Potatoes and Knowledge." Pp. 209–227 in *An Anthropological Critique of Development: The Growth of Ignorance,* ed. Mark Hobart. London: Routledge.

Verran, Helen, and David Turnbull. 1995. "Science and Other Indigenous Knowledge Systems." Pp. 115–139 in *Handbook of Science and Technology Studies,* ed. Sheila Jasanoff, Gerald E. Markle, James C. Petersen and Trevor Pinch. Thousand Oaks, CA: Sage.

Vincent-Lancrin, Stephan. 2006. "What is Changing in Academic Research? Trends and Future Scenarios." *European Journal of Education* 41, no. 2: 169–202.

Wilson, Matthew W., Barbara S. Poore, Francis Harvey, Mei-Po Kwan, David O'Sullivan, Marianna Pavlovskaya, Nadine Schuurman, and Eric Sheppard. 2009. "Theory, Practice, and History in Critical GIS: Reports on an AAG Panel Session." *Cartographica* 44, no. 1: 5–16.

Wynne, Brian. 1996. "Misunderstood Misunderstandings: Social Identities and Public Uptake of Science." Pp. 19–46 in *Misunderstanding Science? The Public Reconstruction of Science and Technology,* ed. Alan Irwin and Brian Wynne. Cambridge: Cambridge University Press.

Zook, Matthew, Mark Graham, Taylor Shelton, and Sean Gorman. 2010. "Volunteered Geographic Information and Crowdsourcing Disaster Relief: A Case Study of the Haitian Earthquake." *World Medical & Health Policy* 2, no. 2: Article 2.

Fisheries Privatization and
the Remaking of Fishery Systems

Courtney Carothers and Catherine Chambers

ABSTRACT: This article draws on directed ethnographic research and a review of literature to explore how the commodification of fishing rights discursively and materially remakes human-marine relationships across diverse regions. It traces the history
of dominant economic theories that promote the privatization of fishing access for
maximizing potential profits. It describes more recent discursive trends that link the
ecological health of the world's oceans and their fisheries to widespread privatization.
Together, these economic and environmental discourses have enrolled a broad set of
increasingly vocal and powerful privatization proponents. The article provides specific examples of how nature-society relationships among people, oceans, and fish are
remade as privatization policies take root in fishery systems. We conclude with an overview of several strategies of resistance. Across the world there is evidence of alternative
discourses, economic logics, and cultures of fishing resistant to privatization processes,
the assumptions that underlie them, and the social transitions they often generate.

KEYWORDS: catch shares, fisheries, individual transferable quotas (ITQs), political ecology, privatization

Nation-states began privatizing fishing rights on a large scale nearly four decades ago. Since that
time, 35 nations have restructured major fisheries, implementing nearly 400 access privatization
programs to manage over 850 species (Environmental Defense Fund [EDF] 2012; Melnychuk
et al. 2011). We use the term *privatization* to signify a variety of processes that redefine access
rights or privileges to open, common, or state-owned fisheries. While true privatization implies
"assigning clear, legally enforceable private property rights to hitherto unowned, state-owned,
or communally owned aspects of the social, cultural, and/or natural worlds" (Castree 2010: 10),
we use the term here to describe many processes that increase the level of private allocation of,
and control over, public resources. Privatization of fishing rights often involves new processes
of *marketization,* creating mechanisms for the monetary exchange or transfer of fishing rights
or privileges between individuals, corporations, or other collectives, and relatedly, *commodification,* reshaping the access rights to fish into objects that can be bought and sold.

There is a range of variation in the nature of privatization, marketization, and commodification processes in various fishery systems worldwide. A consistent step in these diverse privatization processes is the private allocation of resource rights, often in the form of "individual fishing
quotas." These individual fishing quotas usually confer to fishermen, fishing companies, or less
frequently communities or collectives the right to fish for a certain portion of a total catch limit.
Approximately 80 percent of all individual fishing quota programs worldwide allocate fishing

rights as tradable commodities (Bonzon et al. 2010), and are often thus named, individual trans-ferable quotas (ITQs). Among fishermen, scientists, managers, and interested citizens, ITQs have been, and continue to be, deeply polarizing. Agnar Helgason and Gísli Pálsson (1998: 131) note that "the ITQ system has become one of the most contentious and tumultuous issues in Icelandic political history." The US Congress put a moratorium on ITQs for almost a decade in response to widespread concern about the equity issues that result from fisheries privatization. A growing environmental discourse advocating for individual fishing quotas, or "catch shares," for environmental conservation goals has recently reinvigorated privatization processes.

As Petter Holm and Kåre Nolde Nielsen (2007: 193) note, the "ITQ literature is massive." Several syntheses of this burgeoning literature provide helpful reviews of a wide range of ITQ case studies (e.g., Shotton 2000a, 2000b, 2001), the relationship between catch shares and fish resources (e.g., Chu 2009; Costello et al. 2008; Melnychuk et al. 2011), and the social impacts of fisheries privatization (e.g., Copes 1986; Lowe and Carothers 2008; McCay 1995, 2004; Olson 2011; Pálsson and Pétursdóttir 1996). This article, contributing to a volume on capitalism and the environment, does not attempt to provide an exhaustive review of this literature, but focuses instead on exploring how the privatization, marketization, and commodification processes in fishery systems discursively and materially remake human-marine relationships across diverse regions. The article is informed by previous and current directed ethnographic study of fisheries privatization processes in Alaska and Iceland.[1] It traces the history of economic theories that promoted the privatization of fishing access for maximizing potential profits from common property fisheries. It then describes more recent discursive trends that link the ecological health of the world's oceans and their fisheries to the widespread implementation of private property rights. Together, these economic and environmental discourses have enrolled a broad set of increasingly vocal and powerful privatization proponents. Next, it tempers this enthusiasm for privatization with a detailed look at how nature-society relationships among people, oceans, and fish are "remade" (Braun and Castree 1998; Heynen et al. 2007; Mansfield 2008) as privatization policies take root in fishery systems. The article concludes with a presentation of several strategies of resistance to this reframing and remaking of fishery systems.

Fish and Property: Tragedies, Crises, and the Inevitability of Privatization

Increasingly in both academic and popular literature, the fate of the world's fish stocks is linked to the widespread privatization of fishing rights (e.g., Costello et al. 2008; EDF 2012; Festa et al. 2008; Weiss 2008). Environmental Defense Fund, a leading US environmental nongovernmental organization (NGO), has recently created a catch shares design center and manual, including a seven-step process for creating new privatization programs (Bonzon et al. 2010). Its website prominently displays the more than fourfold increase in the number of ITQ programs in effect worldwide since 1990 (EDF 2012). The enthusiasm for fisheries privatization is easily perceivable in scientific literature and conferences. At the International Marine Conservation Congress (of the Society for Conservation Biology) held in Washington DC, in 2009, for example, a speaker introduced his presentation about the benefits of catch shares in fisheries management by showing an image of two small boys drinking a single milkshake with two straws. As he explained, in the absence of separate glasses that would fairly divide their shares, the kids are doomed to compete for the precious milkshake, each trying to drink as much of it as he can before the other. Like many popular stories about shared resources, this story also ends in tragedy. As the presenter explains, rather than enjoying their milkshake at a leisurely pace, both boys end up experiencing a painful ice cream headache. To many conservation scientists in the room that day, the link between two children "competing for" a milkshake and fisheries management needed little explanation. The tragedy of the commons narrative (Gordon 1954; Hardin 1968) has been so overused in describing the problems of fisheries management that today it is not presented as an empirical question but rather as dogma. According to this tale, fishermen—in the absence of private property rights, acting as self-interested, competing, and profit-seeking individuals or businesses—race to outfish their fellow fishermen, securing as much of an open access resource as possible, ending in ruin for all, both the fishermen and the fish.

Why has this tragedy of the commons problem and property-rights-solution framing become such a potent and dominant discourse? Networks of scientists, fishing industry leaders, advocacy groups, and policymakers that have historically, and are currently, promoting catch shares have complex rationales. However, two points appear central to understanding the recent fervor for catch shares. First, the discourses of fisheries privatization, as presented by neoclassical economists, appear as common-sense facts, articulating well with processes of management that strive for objective science informing policy (Wingard 2000). Neoclassical economics provides to capitalist logics the scientific abstraction and mathematical modeling that make these logics appear to be natural, defining features of human society (Davis 1996; Polanyi 1944; Wilk 1996). Theoretical abstraction has made the privatization of fisheries appear as inevitable progress in fisheries management across diverse political processes. Second, the enrollment of a diverse set of actors, promoting much broader goals than aggregate profit maximization, including resource conservation, has made it increasingly common to link fisheries privatization to a host of positive outcomes. Within this broadening of goals has also come a powerful linkage with crisis narratives of overfishing. Scientists, conservationists, and diverse publics are now being told that without enclosure and privatization, fisheries are bound to collapse (e.g., Costello et al. 2008, Weiss 2008).

Economic Science and Capitalist Logic: Privatization for Rent Maximization

> The (economics) profession's most important policy achievement must surely be its influence on getting the ITQs on the agenda as a viable policy instrument.
> —Wilen (2000: 321, cited in Holm and Nielsen 2007: 176)

The tragedy of the commons framing of fisheries was first clearly articulated in scientific literature in the 1950s. Resource economists noted that open access fisheries managed only biologically for total catch limits did not generate any aggregate profits, or resource rents, as did land and other natural resources. If enclosed by private property rights, fisheries could generate maximum profits for firms or profit-seeking individuals that were shared too widely under common property regimes (Anderson 1976). H. Scott Gordon (1954) was among the first to specifically define the absence of private property rights as the key problem of fisheries, and Francis Christy (1973) the first to set out individually allocated "fishermen's quotas" as a solution. Many scholars have since provided critical reviews of this economic framing of fisheries (e.g., Acheson 1981; Carothers 2008; Davis 1996; Macinko and Bromley 2002, 2004; Macinko and Shumann 2008; Mansfield 2004, 2008; McEvoy 1986; Reiser 1999; St. Martin 2005, 2007a, 2007b, 2008), and others have offered a more general critique of such framings that obscure a wide range of successfully managed commons worldwide (e.g., Ciriacy-Wanthrup and Bishop 1975; McCay and Acheson 1987; Ostrom et al. 2002). According to neoclassical economic theory—in the absence of property rights—a "persistent and inevitable" process ensues whereby fishermen work to outcompete each other for their catches of fish, thus investing in "unproductive labor and capital" and "dissipating" all potential resource rents possible from the natural reproductive fish stocks (Moloney and Pearse 1979: 860). The solution is to eliminate the "totally useless accumulation of excess capital" (Crutchfield 1979: 751) and labor (i.e., fishing boats, gear, and fishermen) to enclose the fisheries for fewer individuals and vessels thus maximizing profits for the fleet that remains. The social goal of fisheries according to this economic framing is to maximize aggregate profit for the most efficient fishermen or firms. These goals of efficiency and profit are presented as natural facts of human society, reflecting the close connection between neoclassical economics and capitalism (Polanyi 1944; Wilk 1996). Market mechanisms are seen to be more neutral than democratic processes because they "get the politics out" of fisheries management and provide for the primary goal of fishery systems in a capital-centric imaginary—the "maximization of economic benefit in the long term" (Hannesson 2006: 161).

Since the 1950s, resource economists have developed increasingly sophisticated models for representing optimal allocations of fishing effort to maximize profits. A common framing of fisheries privatization by fishery managers and scientists is that the so-called technical realities of the global crisis of too many boats chasing too few fish demand rationalist management measures such as privatization, often termed "rationalization" by fisheries economists and managers. Because economists tend to use value-neutral language backed by abstract and highly specialized representations, they often more easily align with policy processes that prioritize objectivity and attempt to erase the political dimensions of management. Bourdieu (1999: 165) reminds us that once relegated to a body of specialists, certain discourses can gain a "monopoly of legitimate cultural production." In many regions, the economic discourse of fisheries privatization has gained an ideological and commonsensical power that has obscured its normative framing. This power may, in part, explain the widespread adoption of fisheries privatization policies.

Expanded Goals: Fate of Oceans Linked to Privatization

Strikingly, current proponents of fisheries privatization do not often cite rent maximization as a primary goal. The Environmental Defense Fund (2012), for example, states that catch shares can "bring back fish populations, save commercial fishing jobs, ensure fishing communities prosper and thrive, preserve our fishing heritage, and attract new participants." Conservation of fish stocks is first among these benefits and is increasingly used as a rationale for fisheries privatization in both academic and popular media and discourse. The *New York Times,* the *Los Angeles Times,* and the *Economist* popularized the findings of Christopher Costello, Steven Gaines, and John Lynham (2008) presented in *Science:* "privatization prevents collapse of fish stocks." If ITQs transfer a secure property right to individuals, some argue that those ownership rights will foster a conservation ethic among participants who, as owners, want the fishery resources to remain healthy over the long term for their own benefit (e.g., Hannesson 2005; NOAA Fisheries Service 2012). However, this relationship between privatization and conservation has often been theoretically assumed (e.g., Festa et al. 2008; Fujita and Bonzon 2006), rather than empirically documented (Brandon 2004).

Recent studies have attempted to make up for the lack of investigation of the link between ITQs and increased resource conservation (e.g., Chu 2009; Costello et al. 2008, Heal and Schlenker 2008; Melnychuk et al. 2011), but no simple relationship emerges from this literature. Some authors conclude that ITQs confer benefits to fish stocks (e.g., Costello et al. 2008); others show how stock declines occur years after ITQ implementation (e.g., Copes and Pálsson 2000). Several factors appear to affect the relationship between ITQs and fishery stocks, including the association between total catch limits and division of those limits into ITQs, the property relationship conferred by ITQs, and the fishing practices generated by ITQs. First, the distinctions between various types of quotas are important to consider. Among nonspecialists, the discourse of ITQs may imply that what is being implemented with individual fishing quotas are limits on overall harvests, which have obvious conservation goals. However, the point of ITQs is to divide up (e.g., among individuals, firms, or other collectives) a total, fleetwide catch quota, that is usually set with biological and ecological considerations in mind. Total catch limits are set and enforced in many fisheries managed without ITQs, and have been effective in some ITQ fisheries long before individual quota implementation. A recent study of 345 privatized access fisheries concludes that while catch share programs tend to decrease overexploitation, that "appears to be due more to the presence of a fleet-wide quota cap than to the division of that quota into shares" (Melnychuk et al. 2011). This is a key point. Setting and enforcing total catch limits has obvious benefits for fish stocks. Individually dividing up that total catch limit into tradable commodities has less clear effects on fish stocks.

A second confusing issue that is often glossed over in pro-catch shares discourses is the actual property rights or privileges conferred by ITQs. As Daniel Bromley (2008) notes, there is much "conceptual confusion" about the theory and practice of ITQs. In the economic discourse described above, the closer ITQs are to true resource privatization the better (Hannesson 2006). True privatization—according to these theories—implies stable, clearly defined ownership rights, free transferability, and long-term planning possibilities, thus setting the stage for the increased efficiency and rent maximization, the economic goals that provide the impetus for ITQs in the first place. In the environmental discourse, those who link resource stewardship to ITQs do so under the assumption that real property rights engender a conservation ethic. Oddly, the broader set of proponents of ITQs, including environmental NGOs and fishery managers, tend not to employ the language of privatization. According to Macinko and Bromley: "It is common for IFQ proponents in the United States to reassure a nervous public that they aren't privatizing anything—they are simply advocating what they consider to be the best management tool. Yet proponents then lapse into justifications for IFQs that are thoroughly predicated upon a logic in which privatization is not only beneficial but also necessary" (2002: 21). In the United States, ITQs are defined as revocable access privileges without legal liability, recently described as being conferred in "(revocable) perpetuity" (Abbott et al. 2010). In Iceland, ITQs are permanent shares of a total catch quota, although the Icelandic Supreme Court has ruled ITQs are not equal to private property rights in perpetuity (Hannesson 2006: 78), and a 2012 bill before the Icelandic Parliament aims to define the longevity of ITQ rights. In New Zealand, ITQs are perhaps closest to private property, because they are awarded in (actual) perpetuity; and in the United Kingdom and the Netherlands, long-term rights are not conferred by ITQs (Shotton et al. 2001).

Third, the actual fishing practices employed in ITQ fisheries demand exploration if any conclusions about ITQs and resource health are to be drawn. Parzival Copes (2000) discusses both how ITQ systems in general, and the various specific fishing practices such systems may encourage (e.g., high-grading, or selectively choosing to fill an individual quota with premium fish while discarding suboptimal fish), can create negative impacts on fishery stocks. The practice of leasing quota share from owners by fishermen is common in many ITQ fisheries. Those captains and crew who directly interact with the fish often do not own rights to the resource; thus predictions about how they will or will not behave for the long-term interest of that resource are not straightforward. For example, disenfranchised captains and crew members who resent such leasing practices (e.g., Lazrus et al. 2011) cannot be assumed to employ fishing practices with long-term resource sustainability in mind. The rural-to-urban migration of ITQs also demands attention. Those individuals who reside in coastal communities, who may well be more concerned about the long-term health of their marine ecosystems than those nonlocals who migrate in and out only for commercial fishing, are often dispossessed of fishing rights after resource privatization (Carothers 2012). The diversity of practices catalyzed by ITQs demands attention before broad generalizations about ITQs and resource outcomes are made.

Each of these points demonstrates the unclear relationship between privatization and resource conservation. Despite these ambiguous relationships, the discursive linking of fisheries privatization to environmental stewardship and ecosystem health (as well as other goals, such as increased human safety[2]) has had important implications. In the language of actor network theory (Callon 1986; Latour 1987), the expanded goals of fisheries access privatization have successfully enrolled important actors in fisheries and marine management and conservation networks. Within this expanded environmental discourse, resistance to privatization is increasingly imagined not as resistance to dominant cultural logics that promote the marketization and commodification of fishing rights for maximizing profit, but to the environmental logics that are redefining a new environmentalism of marine ecosystems. This emerging environmental discourse has constructed privatization processes, both of access rights to extractive fisheries and of marine spaces, as necessary precursors for protecting ecosystem integrity and health. These discourses have mobilized environmental activists, marine recreationalists, coastal tourists, and seafood consumers to support privatization processes. ITQs have now been reframed as "catch shares." The individual and tradable nature of the resource rights that defined the economic framing is no longer central to this environmental discourse, in which catch shares are the "new hope for fisheries" and a "real investment in sustainability" (EDF 2012: n.p.).

From Inevitable Natural Logics to Capitalist Production Systems

The dominance of the tragedy of the commons metaphor in both economic and environmental discourses, paired with the requisite privatization of access rights to fishing, has gained such commonsense status in international fisheries science and management that it is often discussed in terms of its inevitability (Árnason 1993; Hannesson 2006). For example, a fishery manager in Alaska described the push for fisheries privatization, often called "rationalization" in this region, in Kodiak in March 2006: "We had a situation, worldwide, where people in fisheries are having an increasing race for fish … more and more boats, more and more people, more and more technology and gear going after the same number of fish. And worldwide there has been a lot of variation, but there's been one form or another of rationalization." In addition to articulating the perceived inevitability of fisheries management regimes moving toward privatizing fisheries, this statement presents the problems of fisheries as a technical one of too much steel and too few fish. Missing here, and often in both economic and environmental discourses, is the key point that this "race for fish" is not a technical problem created by individual, unconnected human behavior presented in the tragedy of the commons story, but rather a political economic problem created by larger-scale processes of global capitalism and industrial fisheries.

Utilizing a political ecology framework, Becky Mansfield (2010) provides an overview of both the critique of tragedy of the commons framings and the industrial fisheries development that has led to global fisheries crises. As she notes, overfishing is not caused by the human propensity for individual greed, but rather industrial processes that have modernized and developed fisheries in recent decades. The scale of industrial fisheries became massive in the 1960s and 1970s; technological development enabled fishing fleets in the US, UK, Spain, Japan, and Russia, to travel farther and catch and process more fish (McGoodwin 1990). During this development, Mansfield also notes the inequitable flow of resource wealth of fisheries from developing to developed countries. Nearly three-quarters of all fish traded internationally comes from nations in the Global South exported for consumption in nations in the Global North, over 70 percent of which is for markets in the European Union, Japan, and the United States. Overfishing in the

Global South is caused in large part by export commodity markets and the demand generated in these wealthier countries (Mansfield 2010).

Rather than an inevitable process, we see industrial fisheries developing for specific reasons. Fisheries were targeted as an underdeveloped realm for increased economic growth. In the United States, federal policies aimed at developing fishing capacity in the 1960s and 1970s resulted in a near doubling of commercial fishing vessels in the country between the mid-1960s and 1980s, accounting for two-and-a-half times more fish harvested between the 1970s and 1990s (Wingard 2000). Large-scale subsidies financed between 50 and 87 percent of costs for constructing or refurbishing fishing vessels, shore-based infrastructure, and marketing (Mansfield 2010; Wingard 2000). Today, approximately US$16 billion are spent on increasing fishing capacity worldwide, and US$4–US$8 billion are spent on fuel subsidies (Mansfield 2010).

Rather than situate overfishing in a global system of uneven development that generates large-scale extraction, cheap products, inequitable flows of resource wealth from south to north (Mansfield 2010), both economic and environmental narratives tend to link overfishing to inevitable human nature (tragedy of the commons). These discourses, paired with the striking statistics of the overfishing that has resulted from the industrialization of fisheries, where 32 percent of all stocks are estimated to be "overexploited, depleted, or recovering," 50 percent are fully exploited, 12 percent moderately exploited, and 3 percent underexploited (Food and Agriculture Organization [FAO] 2010: 35), have created a fertile context for fisheries restructuring. Yet, the overfishing problem constructed in this way as one of individual human greed, paired with a solution of privatization, leaves unimagined alternatives that would actually address the root causes of overfishing—the industrialization and uneven development of fishery systems.

Remaking Nature-Society Relationships

Social scientists have increasingly explored relationships between the environment and privatization (e.g., Heynen et al. 2007; McCarthy and Prudham 2004). Building on the work of Karl Marx, scholars have explored how privatization processes and the new property relationships they create often dispossess people from land, sea, and resources, such that they are forced to labor for the owners of capital and newly created commodities (Mansfield 2010). Wealth is often accumulated through these processes of dispossession (Harvey 2003). In the fisheries context, the connection between the economic theories that promote the privatization of fisheries access and wealth accumulation is explicit. The point of resource privatization—as described in economics literature from the 1950s to the present—is to maximize potential profits, or resource rents, by eliminating so-called wasteful labor and capital. Processes of privatization are often conceived of as inevitable processes for those employing capitalist logics, so alternatives are left unimagined in this economic literature (Carothers 2008; Davis 1996; St. Martin 2007a), except in cases of perceived extreme difference, such as indigenous fisheries or fisheries in isolated communities, where more attention might be paid to fishing as a way-of-life rather than as a profit-generating endeavor (e.g., Crutchfield 1979: 751). The environmental logics that promote the privatization of resource rights and marine space for preserving marine ecosystems also generate dispossessions. For example, prioritizing uses of marine spaces for scientific research and maritime recreation in marine reserves over extractive fishing redefines legitimate uses of marine resources and spaces and often reframes fishing as a practice that threatens ecosystem structure and function.

The question we explore in this section is how privatization processes remake nature-society relationships in fishery systems. Scholars like Bruce Braun and Noel Castree (1999), Becky Mansfield (2004, 2008, 2010), and Julia Olson (2011), who write about the ways in which enclosure and privatization processes remake nature-society relationships, are careful to caution against broad-brushed generalizations. As many case studies of fisheries privatization in diverse contexts reveal, places, peoples and natures vary considerably, as do the specific impacts of restricting, commodifying, and marketizing fishing rights. Yet some general trends are observable when comparing processes of fisheries access privatization across the globe. Olson presents an impressive review of the social impacts of fisheries privatization and the "new ways of organizing around fishing" (2011: 353) that emerge once privatization policies are enacted. Olson documents cases in the US, Canada, Iceland, Norway, and Australia, describing the diverse impacts of privatization based on national approach, fishery, and social context; however, a key finding presented is that "negative impacts from privatization often fall on less powerful segments of the fishing industry, namely the crew, or the small business owners" (Ibid.: 361). Those best able to reap the benefits of new programs, larger firms or vertically integrated corporations, come to redefine the structure of fisheries—by leasing out rights and shifting hiring and compensation practices for captains and crews, and of fishing communities and regions that are made up of those captains and crews among other fishery participants (Olson 2011).

Initial Allocation, Market Trading, and Concentration of Wealth and Power

Several scholars use the concept of experimentation to describe the implementation of ITQs (e.g., Copes and Pálsson 2000; Helgason and Pálsson 1998; Olson 2011; Reiser 1999). Resource economists and marine scientists often view nature as something essentially manageable, and ITQs are a direct extension of this notion (Pálsson and Helgason 1996). Common outcomes of fisheries privatization are fully predicted to occur according to the economic theories that promote it, including the concentration of wealth derived from fishing. Among the first steps in privatization processes is allocating the newly defined fishing rights. ITQs have been most commonly allocated to boat owners (individuals, firms, or in some fisheries, vessels), even in fisheries where hired skippers are common. This common practice is based on a capital-centric logic of rewarding those who have invested financially in fisheries, rather than with their labor; "vessel owners and lease holders are the participants who supply the means to harvest fish, suffer the financial and liability risks to do so, and direct the fishing operations" (NOAA Fisheries Service 1993: 59378). An exception is the Bering Sea crab fishery privatized in 2005 that allocated hired captains some fishing quota, although collectively they received only 3 percent of the total fishing quota. In many other fisheries, hired skippers and crew received no portion of the newly created wealth, often generating much ill will between boat owners and operators. These tensions contributed to an anti-ITQ movement among fishermen in the US, resulting in a seven-year moratorium on ITQs (McCay 2001; see also Criddle and Macinko 2000).

The proportion of the total catch allocated to owners or vessels in an ITQ system is usually based on past catch levels in the fishery. These historic catches often reflect an average catch of a vessel over a short range of years (Shotton 2001). This pattern of allocating fishing rights based on the percentages of catch within a set number of years has catalyzed a phenomenon where fishermen fish for "fishing history" in fisheries they expect to become privatized (Copes and Pálsson 2000; Maurstad 2000). In the Alaska halibut fishery, for example, Clarence Pautkze and Chris Oliver (1997: n.p.) note that "effort pour[ed] in to establish a fishing history and rights to what probably w[ould] be a permit of considerable value." In this fishery, as in many others, those boats that caught the most fish and most contributed to the problems of overcapacity in need of restriction were rewarded with the most fishing rights. Bonnie McCay (2001) notes that even the "catch history" for illegally harvested clams was used to determine future fishing rights in the Mid-Atlantic region of the United States.

Economic theories predict that those who are most efficient (e.g., have lower costs and make higher profits) will buy out those less efficient fishermen (see, e.g., Árnason 1993; Crutchfield 1979; Hannesson 2006). The market mechanism creates a space that favors those who are economically efficient and have access to capital (Eythórsson 1996). Those most efficient with access to capital are often large, vertically integrated firms (Copes and Charles 2004; McCay et al. 1995; Pálsson and Helgason 1995). In Iceland, consolidation of quotas occurred quickly in the first four years after ITQ implementation (Pálsson and Helgason 1995). In 2009, the 5 largest fishing companies held 33 percent of all fishing quota and the 20 largest held 66 percent (Icelandic Ministry of Fisheries 2011). Over the course of a decade, ITQs held in New Zealand fisheries consolidated substantially. By 1996, 86 percent of total catch quotas were owned by the 12 largest companies (Stewart et al. 2006). Individual ownership limits in New Zealand are currently among the highest worldwide. An individual firm may own as much as 35 to 45 percent of total quota in a fishery, limits that do not allow an outright monopoly but provide for an oligopoly of control by a few companies (Stewart and Callagher 2011). Dramatic consolidation has occurred in some ITQ fisheries in the US. For example, the fleet of vessels participating in the Bering Sea red king crab and snow crab fisheries in 2004 shrunk by 63 percent and 58 percent respectively during the first full year of ITQ management (2005 and 2006 respectively); currently, both fleets number just over 30 percent of their pre-ITQs highs (Abbott et al. 2010). In some fisheries, limits on consolidation and fishing quota linked to vessel-size classes have prevented widespread wealth concentration (Fina 2011; National Research Council [NRC] 1999a).

Fleet consolidation often translates into a direct loss of jobs, especially for nonquota owning captains and crew. Job loss can be rapid, as in the Bering Sea crab fisheries where approximately 70 percent of crew jobs were eliminated with the privatization of fisheries access rights (Abbott et al. 2010). Economists assume those less efficient quota shareholders are properly compensated because they sell their fishing quota, and moreover, they better serve society because their labor is assumed to be mobile and can be better put toward other ends (Hannesson 2004). However, several case studies take issue with these conclusions, noting instead the role that culture and economic stress can play in bringing about the sale of fishing rights by those who highly value them (e.g., Carothers 2010, 2012), and that labor and capital based in communities is often not mobile (Davis 1996). Gunnar Knapp (2011) recently demonstrated in Alaska salmon fisheries that as the value of a fishery increases, the level of local ownership decreases. Low-income and indigenous fishermen tend to be more likely to sell ITQs in the Alaska halibut fishery and less likely to purchase them (Carothers in press; Carothers et al. 2010). These trends have led to a rural-to-urban migration of fishing quota out of small (under 1,500 people), remote coastal communities in the Gulf of Alaska region; the number of rural fishermen in this region that hold halibut quota has decreased by 40 percent and the number of rural halibut quota share holdings has decreased by 56 percent since 1995 (NOAA Fisheries Service 2010). In Iceland, a similar rural-to-urban shift in fishing ownership and operation has occurred. Vertical integration of fishing companies and the increasing scale of fishing vessels and processing plants has limited employment options for fishermen and processing workers in rural communities.

Olson (2011) notes privatization is not always accompanied by consolidation or an increasing concentration of wealth. In the case of Australian South East Trawl fishery privatized during a time of wide unemployment and uncertainty about future fishing prospects, the number of fishers remained fairly constant (Connor and Alden 2001). New Zealand fleets expanded after the implementation of ITQs largely due to state support of offshore fishing capacity (Olson 2011).

If ITQs lengthen fishery seasons, other fishery sectors may increase employment opportunities, for example in the processing and support business sectors (Olson 2011). Contrary to these common trends of the concentration of fishing rights more inequitably in increasingly fewer hands, individually allocated fishing rights can also be used as equalizers of wealth. The Norwegian implementation of ITQs in 1990, for example, provided for equivalent incomes for boat-owning fishermen; however, equity issues were created for those without boats (Olson 1997, 2011).

Changes in Fishing Relations and Practices

> Some of these quota lords are friends of mine. And they kind of know the deal, but they're caught in this whole capitalist thing … I've given up on the halibut and the black cod fishery. It's just a lost fishery to me at this point. You know, chasing – allowing myself to get Q-teased.
> —Interview with a Kodiak fisherman (July 2011)

Kevin St. Martin (e.g., 2005, 2007a, 2007b, 2008) has explored processes of capitalism and "noncapitalism" in New England fisheries. He explores how the use of a share, or lay, system of compensation as distinct from wages is characteristic of alternative economic arrangements that characterize many fishing systems. Fishing crew members, like captains and owner-operators are usually compensated for their labor in catching fish (and other vessel and shore-based work) by sharing in the earnings of the catch (minus some shared expenses, commonly including groceries, fuel, and bait) rather than a per hour, day, or trip wage (McCay 1995; St. Martin 2008). In many cases, the privatization of fishing access has restructured similar share systems, often dramatically reducing the shares paid to crew. Prior to the implementation of ITQs in the US North Pacific halibut fishery, for example, crew members collectively earned a share of the gross profits up to 70 percent. After the implementation of halibut ITQs, this crew percentage has dropped to 33 percent (Rosvold 2007). Though economic analyses of these shifts point out that crew earnings rose with the increased value of halibut following the introduction of ITQs (Hartley and Fina 2001; Herrmann and Criddle 2006), a much larger percentage of the rent generated by access privatization went to quota owners. Prior to the privatization of the Bering Sea crab fisheries in 2005, crew members collectively tended to receive approximately 35 percent of gross vessel revenues; this percentage dropped to 23 percent by 2010 (Fina 2011). Public testimony given by crab crew members suggest that some vessels in these fisheries maintain historic crew percentages, but other boats have dropped their individual crew shares from 5–6 percent to 0.5–2 percent after ITQ implementation (Dochtermann, personal communication, 2012). In these fisheries, Abbott and colleagues (2010: 333) find that "both seasonal and daily employment remuneration increased substantially for many crew" after the privatization of the fishery, although earnings per crab caught decreased "as a result of increased crew productivity and the necessity of paying for fishing quota in the new system." Evelyn Pinkerton and Danielle Edwards (2009) note a shift in crew shares from 10–20 percent in Canada to 1–5 percent post-ITQ implementation in the British Columbia halibut fishery. Although the value of that fishery has increased about 25 percent over the last 20 years, the proportion of value of the fishery going to crew members has dropped by over 70 percent.

One primary rationale for the decreasing shares given to crew members in privatized fisheries is the widespread practice of quota leasing and deducting ITQ "costs" from vessel earnings before crew shares are paid. Pinkerton and Edwards (2009) refer to quota leasing in the British Columbia halibut fishery as the elephant in the room because nearly 80 percent of all quota in this fishery is leased, yet these practices have received relatively little analysis. Lease prices doubled between 1993 and 2008, reaching 78 percent of the total value of the catch (Pinkerton and Edwards 2009). In the Bristol Bay red king crab fishery, lease fees can reach approximately 70 percent (Lazrus et al. 2011). Deductions for the ITQ costs have risen in this fishery from 40 percent in 2005 to 80 percent in 2006 and 2007; approximately 80 percent of vessels deduct ITQ costs from crew shares (Abbott et al. 2010), although the practice of charging for initially allocated "owned quota," or the "opportunity costs" of ITQs is still considered unfair by many quota owners (Abbott et al. 2010; Lazrus et al. 2011). In the halibut fishery, rates of ITQ deductions have been one of "decelerating acceleration," as one Kodiak fisherman noted. "At first people weren't charging anything, then it was 20 percent, 50 percent, 60 percent, 65 percent. So it went up pretty fast in the first few years. When people started buying it then they had debt service, so they're financing it with a piece off the top" (interview with a Kodiak fisherman, April 2011).

In Iceland, as quota accumulated in the hands of larger companies, small-scale fishermen were often forced to lease quota from them (Helgason and Pálsson 1997; Pálsson and Helgason 1995, 1996). Fishermen were placed under the direct control of the companies in terms of wages earned. In Icelandic fishing discourse, fishermen became "tenants" who were "fishing for others" under control of the "lords of the sea." In British Columbia, quota owners often charge their lease fees through middlemen like processing plants because of the ethical issues they have with charging such fees directly to fishermen (Pinkerton and Edwards 2009). Some fisheries privatization programs, such as those in Australia do not prohibit outside investors from buying quota share; this practice further removes fishermen from ownership (Bradshaw 2004; Copes and Charles 2004; Pálsson and Helgason 1995). Helgason and Pálsson (1997) suggest that the increase in the prevalence of quota share leasing marks a shift in the nature of fishing rights such as individual transferable quotas. This practice of profit making from fishing rights without fishing shows that ITQs are "not just use-rights, but, in effect, property rights that could be exchanged solely for profit" (Helgason and Pálsson 1997: 457; see also Karlsdóttir 2008).

In several ITQ fisheries the crew share system has entirely shifted to a wage system (e.g., British Columbia halibut and the Tasmanian rock lobster fishery [Olson 2011]). Recent interviews in Kodiak suggest that some captains in the Alaska halibut and other ITQ fisheries are employing more wage workers as fishermen operating outside of the customary share system. These workers, commonly called "$100/day guys," are often viewed negatively by long-term crew members used to being part of a share system they see as being eroded by the new restructuring that accompanies access privatization. This practice reflects to some crew members a devaluing of their knowledge and skills.

These shifts in compensation practices, while not universally negative for all crew in all ITQ fisheries, do further institutionalize and legitimate uneven distributions of wealth that flow from fisheries, particularly resource rent created by privatization processes. In many ITQ fisheries the value of the ITQs reflects real costs for operators who have purchased or leased quotas, or opportunity costs for those who were originally allocated quotas. The owners of the capital have realized increased benefit and power, while the labor class has become less able to obtain ownership of their own means of production. The changing nature of fishing relationships in many ITQ fisheries has substantially decreased upward labor mobility, often creating impassable class divisions. In many ITQ fisheries, it is not a realistic expectation that a fishing crew member could become a fishing captain or boat and quota owner. The added capitalization of fishing rights has severely eroded options for crew members or new entrants who cannot afford them (e.g., Lowe and Carothers 2008; Yandle and Dewees 2008).

The further entrenching of class roles has shifted the power relationships in fishing. In Alaska ITQ fisheries, long-term crew members describe being "Q-teased." Crew may dedicate boat and gear labor time for the promise of a crew position on a future ITQ trip that may not materialize, or may be offered work as a crew member in other less lucrative fisheries with the potential for ITQ trips in the future. For some long-term crew members, new feelings of dependency on "quota lords" for employment, and uncertain employment, is not worth the promise of potential earnings. In a July 2011 interview, one Kodiak fisherman described this in terms of power shifts: "you no longer have any power as a crewman … the power shifted away … from the people that actually do the work."

Changes in Communities, Cultures, and Economies

> The smaller operators and, it seems, the majority of Icelanders vehemently object to the new commodity identity of fishing rights. The transformation of the social identity of fishing right, from the status of common property to commodity, has turned out to be one of the most contentious issues associated with ITQs in Iceland.
>
> —Helgason and Pálsson (1998: 129)

Fisheries privatization policies remake the social and ecological relationships in fishing. The shifts generated by resource privatization have wider impacts on communities, cultures, and economies. The pattern of rural-to-urban migration of fishing rights is linked to changes in rural coastal economies and cultures. Vertical integration and the increasing scale of fishing in Iceland and other nations contribute to rural depopulation. For example, women in rural coastal communities in Iceland, who had once relied on steady income from fish processing plants are negatively impacted, as are crew and youth who may not easily transition into reproducing the social life of fishing communities (Karlsdóttir 2008; Skaptadóttir 2000). Anna Karlsdóttir (2008) discusses how such policies can influence the ways in which people define their identities and social roles that can have far-reaching social impacts, potentially damaging the moral fabric of communities.

Because ITQ fisheries redefine successful fishermen—leading to professionalization of the fleet, full-time engagements, and a favoring of those with access to capital—the structure and make-up of coastal communities can shift. Empirical studies of the impacts of ITQs suggest that the values of fishing change when quotas are implemented. Anita Maurstad (2000) describes the important distinction between "capitalistic" fishermen, privileged by ITQ management, and other small-boat fishermen who may only fish when they need income. Similarly Anthony Davis (1996) describes the varying motivations for fishing contrasting those of "accumulation-" versus "livelihood-" focused fishermen. In small indigenous communities in Alaska, there is evidence that individualization and competitiveness become more common after fishing rights become commodified and marketized, and "lifestyle" fishermen who do not seek profits from fishing become marginalized in these systems (Carothers 2008, 2012). These shifts in values and relationships often transform community structure as in the example of ITQs in New Zealand fostering new competitive relationships in opposition to traditional Maori social relationships (McCormack 2010).

For indigenous cultures, ITQs can represent another alienation process in a long history of colonialism. For example, Maori participation in commercial fishing in New Zealand prior to ITQs was primarily a part-time, but vital, source of income (Memon and Cullen 1992). The 1992 Fisheries Claims Act defined the commercial and customary use rights of Maori peoples as separate legal, economic and cultural categories. This dichotomization did not accurately reflect the reality of contemporary indigenous fishing cultures and economies (McCormack 2007, 2010) that employed mixed motivations (i.e., fishing for food, health and healing, cultural practice, education, income) for fishing participation. The reduction of fishing practices to either commercial or customary can mean a loss of authority over livelihood and work decisions, a decreased affiliation and identification with ancestral resources, and may undermine attachments to place (McCormack 2007, 2010). Similarly in Alaska, subsistence-based communities that mix commercial and subsistence productive systems have often been constrained by privatized fishing rights (Carothers 2010; Reedy-Maschner 2010).

Resisting the Reframing and Remaking of Fishery Systems

> If we do not want to see our marine resources and related communities, cultures, and jobs disappear, then we would do well to learn from our past lessons.
> —Fujita and Bonzon (2006: 312)

Processes of commodification and marketization that enable massive wealth transfer encourage accumulation by dispossession (Harvey 2003). Many fishermen and fishing communities have been alienated from fishing livelihoods through processes of commodification and marketization of fishery access rights. However, these processes have also generated much resistance, both discursively—resisting the dominant ideas about privatization that have come to frame fishery systems by creating and circulating challenging discourses, and materially—in some instances securing noncommodified and nonalienable access rights to marine-based livelihoods.

Increasingly, environmental, economic, and management discourses of privatization gain power in their ability to define away alternatives. As a result, the ability to articulate and enroll supporters for other framings of fisheries problems and goals is further constrained (Bourdieu 1998). As dominant discourses gain a monopoly on the power to name and define, those marginalized by the processes of enclosure become increasingly portrayed as irrational users (e.g., livelihood fishermen) or unproductive, redundant, expendable labor (e.g., skilled crew members). However, those irrational and redundant users have been vocal in this resistance to fisheries privatization.

Those affected by ITQs have reframed discourse. Fishermen and communities across the globe use explicit expressions of values of fairness and equity to critique the remaking of human-marine relations resulting from privatization processes. Helgason and Pálsson (1997), for example, explore how fishermen and others placed the privatization of their public fisheries resource in a moral language of theft, profiteering, and feudal inequality. In the United States and Canada, the development of a diverse but collective voice of resistance is repoliticizing privatization discourses and processes (Butler 2008; Carothers 2008).

Those affected by ITQs have also taken legal action. In New Zealand, for example, ITQs promoted resolution of Maori customary and commercial fishing rights (Bourassa and Strong 2000; Memon and Kirk 2011). As the eventual result of the 2004 Maori Fisheries Act, all 57 Maori *iwi* (tribal entities) own quota in perpetuity, totaling over 50 percent of the total quota (Dewees 2008). However, these rights to quota are not without difficulty, as this comanagement still operates under an ITQ framework (Batstone and Sharp 1999), and there are still issues regarding access to fish for rural employment. Many *iwi* lease their quota, so it is still hard for Maori to benefit from employment in fishing and fish processing (Memon and Kirk 2011). Dewees (1997: 105) notes that although the Maori now control a large proportion of fishing rights in New Zealand, many Maori informants were of the "general opinion that ITQs were not compatible with their beliefs about fisheries management … due to the Maori emphasis on collective rather than individual focus for distribution of benefits."

Opposition and resistance to ITQs has led to legal action in Canada, Iceland, Norway, and the United States among other nations. Historical use, continued dependence, and shift in ownership resulting from ITQs were factors that influenced the Supreme Court of Canada to rule that the government must provide First Nations "improved access to fishery resources" (Copes and Pálsson 2000: 3). The Sami Parliament contested the implementation of vessel quotas in Norway in 1990 and continues to fight for indigenous rights to fish (Davis and Jentoft 2001; Jentoft and Karlsen 1996). Similarly in the U.S., indigenous fishing tribes in Alaska have used the courts to argue that ITQs violate aboriginal rights to fish (Carothers 2011). The United National Human Rights Committee ruled in late 2007 that the existing Icelandic fisheries law based on ITQs violated the International Covenant on Civil and Political Rights in Iceland. The case, based on an act of resistance in 2001 by two fishermen fishing without quota, had also been tried in the Icelandic Supreme Court and, like similar cases that came before it, questioned the constitutionality of the ITQ system with regards to discrimination and the right to work (Einarsson 2011). "ITQ provisions have discriminated in favor of particular groups in the fishery, allowing them substantial wealth benefits from the public fishery resource, at the expense of others" (Copes and Pálsson 2000: 3).

The emerging disparity between theories guiding ITQ implementation and on-the-ground realities of fisheries and fishing livelihoods has led to several instances of changes made after implementation to ameliorate negative effects of quota system, oftentimes specifically for the purpose of addressing considerations of access and equity. In partial response to contested fishing rights and quota consolidation in Iceland, a separate small boat quota system was implemented for those who fish only with handlines or longlines. In addition, in 2009, the Icelandic Ministry of Fisheries instituted a "coastal fisheries" option, in which fishers are allowed to catch a certain amount of the total allowable catch (totaling 6,000 tons in 2010) without quota, under certain restrictions such as season, a daily weight limit, gear, and boat size. Although there are examples from Iceland of towns adapting to loss of fishing rights by reoutfitting boats for tourism operations (Einarsson 2009), and women coping with loss of processing plant jobs by creating handicraft business (Skaptadóttir 2000), "coastal fishing" offers hope of fisheries reengagement to places with limited employment and is deemed successful by many in small coastal communities that had been drained of quota (Einarsson 2011). New Zealand implemented an annual catch entitlement in 2001 that allows for fishing inshore without needing to own quota, under certain restrictions (Stewart and Callagher 2011). The United States implemented a community purchase option for small communities in Alaska that were otherwise unable to access fishing rights (Carothers 2011; Langdon 2008).

Bonnie McCay (2004) concludes that resistance to ITQ governance has brought about adaptations to these programs that can enhance community-based fisheries management. Property mechanisms can make secure the resource rights of indigenous groups like the Maori and Canadian First Nations. Similarly, the Community Development Quota program implemented in western Alaska in 1992 utilized a quota system to allocate a portion of fishing rights to small indigenous communities who did not participate in the industrial fisheries of the Bering Sea (Mansfield 2008; NRC 1999b). It is clear that countries once considered leaders in quota management design are making changes to alleviate negative aspects of these privatization programs, both in response to legal challenges and to provide alternative opportunities for rural and indigenous fisheries that have been largely constrained by privatization. The necessity of these amendments should lend cautionary notes to recent enthusiasm for continued privatization of fishing rights.

Concluding Thoughts

Fishery systems are complex. Simplistic stories about the tragedy of the commons and kids and milkshakes are not appropriate metaphors to guide thinking about these systems. Many cultures, communities, and economies continue to depend on access to fishery resources, yet across the globe privatization of fishery systems is remaking these systems in ways that often inhibit alternative economies and cultural logics. Fishery system sustainability depends on sustaining cultural and economic pluralism and securing resource rights for coastal residents (McGoodwin 1990). This article has explored how the privatization of fisheries access rights discursively and materially remakes human-marine relationships across diverse regions. The economic theories, capitalist cultural logics, and large-scale experimentations that have led to the privatization of fishing access for maximizing potential profits from common property fisheries now has a forty-year history to review. Many of the common outcomes of privatization are expected: consolidation of fishing fleets, concentration of wealth, increasing efficiencies of vessels and vertically integrated firms, and the rural-to-urban migration of fishing rights. Others such as the solidification of classes of fishing participants, shifts in the relative earnings and power of these classes, changes in fishing practices, and transitions in fishing communities are perhaps less perceptible without detailed study. The economics literature that first promoted these policies had a narrow goal—maximize aggregate potential profits. Privatization of fishing rights for the economic goal of maximizing profits has been shown to often remake fishery systems in the image of capital production, to the detriment of alternative economic and cultural logics that typify a diversity of fishing communities worldwide currently struggling to resist these changes.

Environmental logics that increasingly redefine rationales for the privatization of fishing rights (and related marine spaces) for resource conservation and ecosystem stewardship, as well as other social goals, have enrolled a broader network of supporters who espouse a diversity of values (e.g., Ecotrust 2011). This diversity opens up new opportunities for resistance.[3] Some fishers and fishing communities are forming networks to better resist capitalist restructuring and facilitate pluralism (e.g., Community Fisheries Network 2012). Privatization processes are being reimagined in environmental and community discourses to accommodate goals other than maximized wealth generation. Marketizing and commodifying fishing rights, central for profit generation goals, are being challenged as more people advocate for the inalienability of resource rights and the inherent right to fishing livelihoods. How these new environmental and social logics are redefining privatization processes and reshaping fishery systems warrants more attention.

▪ **ACKNOWLEDGMENTS**

This work was funded in part by the National Science Foundation (Arctic Sciences, Grant No. 1023619) and Fulbright and Leifur Eiríksson Scholarships. We thank several Kodiak fishermen who were interviewed as part of the "Social Transitions in Kodiak Fisheries" project and informed the development of this article. We are grateful for space provided by the University of Alaska Fairbanks Kodiak Seafood and Marine Science Center and the Blönduós Academic Center. We also want to thank the three anonymous reviews for their helpful comments. All errors or misrepresentations are solely our own.

▪ **COURTNEY CAROTHERS** is an assistant professor of fisheries at the University of Alaska Fairbanks. She is an environmental anthropologist whose research program focuses on understanding social, cultural, and economic diversity in fishing communities and explores ways to sustain that diversity into the future. In one central area of study, she explores the material, social, and symbolic shifts in fishing livelihoods as fishing rights become privatized. In another, she partners with indigenous communities in the Arctic to study social-ecological change and subsistence ways of life. Her specific areas of expertise include: political ecology; resource enclosure and privatization; indigenous knowledge, science studies, and politics of knowledge; subsistence, mixed, and alternative economies; socio-ecological change; fishery systems; and Alaska Native cultures.

CATHERINE CHAMBERS is a doctoral fellow in the Marine Ecosystem Sustainability in the Arctic and Subarctic program at the University of Alaska Fairbanks. As a recipient of Fulbright and Leifur Eiríksson scholarships, she is currently conducting research through Hólar University College and the Blönduós Academic Center in Blönduós, Iceland. Her dissertation research focuses on subarctic coastal communities and issues of access and participation in marine-based livelihoods.

◼ NOTES

1. Previous research is summarized in Carothers (2008, 2010, 2011, 2012). Both authors are currently involved in an ethnographic study of experiences of fisheries privatization and impacts on individual and community well-being in Kodiak, Alaska (National Science Foundation, Arctic Sciences, Grant No. 1023619, 2010–2013). Current ethnographic research exploring fisheries privatization processes, among other changes in fishing communities, is also being conducted in Iceland (Fulbright Scholarship, 2011–2012; Leifur Eiríksson Scholarship, 2011–2012).
2. Safety is also becoming linked to ITQ management. When ITQ systems replace derby fisheries, safety improvements often result (see Fina 2011; Hartley and Fina 2001); however, safety gains are realized because of the elimination of unsafe fishery practices rather than the implementation of ITQ programs as such (see Lazrus et al. 2011 for a diverse range of perspectives on safety and ITQ implementation in the Bering Sea and Aleutian Island crab fisheries).
3. We thank one of our anonymous reviewers for making this important point.

◼ REFERENCES

Abbott, Joshua K., Brian Garber-Yonts, and James E. Wilen. 2010. "Employment and Remuneration Effects of IFQs in the Bering Sea/Aleutian Islands Crab Fisheries." *Marine Resource Economics* 25, no. 4: 333–354.

Acheson, James. 1981. "Anthropology of Fishing." *Annual Review of Anthropology* 10: 275–316.

Anderson, Lee. 1976. "The Relationship Between Firm and Fishery in Common Property Fisheries." *Land Economics* 52, no. 2: 180–191.

Árnason, Ragnar. 1993. "Ocean Fisheries Management: Recent International Developments." *Marine Policy* 17, no. 5: 334–339.

Batstone, Chris J., and Basil M. H. Sharp. 1999. "New Zealand's Quota Management System: The First Ten Years." *Marine Policy* 23, no. 2: 177–190.

Bonzon, Kate, Karly McIlwain, C. Kent Strauss, and Tonya Van Leuvan. 2010. "Catch Shares Design Manual: A Guide for Managers and Fishermen." Environmental Defense Fund. http://www.edf.org/sites/default/files/catch-share-design-manual.pdf (accessed 5 January 2012).

Bourassa, Steven C., and Ann Louise Strong. 2000. "Restitution of Fishing Rights to Maori: Representation, Social Justice and Community Development." *Asia Pacific Viewpoint* 41, no. 2: 155–175.

Bourdieu, Pierre. 1998. "The Essence of Neoliberalism: What Is Neoliberalism? A Programme for Destroying Collective Structures Which May Impede the Pure Market Logic." *Le Monde Diplomatique*. http://www.mondediplo.com/1998/12/08bourdieu (accessed 5 January 2012).

Bourdieu, Pierre. 1999. *Language and Symbolic Power.* Cambridge, MA: Harvard University Press.

Bradshaw, Matt. 2004. "The Market, Marx and Sustainability in a Fishery." *Antipode* 36, no. 1: 66–85.

Brandon, Heather. 2004. "Theoretical and Actual Biological Effects of Share-Based Fishery Management Programs: Three Case Studies from the North Pacific." MA thesis, University of Washington.

Braun, Bruce, and Noel Castree, eds. 1998. *Remaking Reality: Nature at the Millennium.* New York: Routledge.

Bromley, Daniel. 2008. "The Crisis in Ocean Governance: Conceptual Confusion, Spurious Economics, Political Indifference." *MAST* 6, no. 2: 7–22.

Butler, Caroline. 2008. "Paper Fish: The Transformation of the Salmon Fisheries of British Columbia." Pp. 75–98 in *Enclosing the Fisheries: People, Places, and Power: American Fisheries Society Symposium 68,* ed. M. Lowe and C. Carothers. Bethesda, MD: American Fisheries Society.

Callon, Michel. 1986. "Some Elements of a Sociology of Translation: Domestication of the Scallops and the Fishermen in St. Brieuc Bay." Pp. 196–233 in *Power, Action and Belief: A New Sociology of Knowledge,* ed. J. Law. London: Routledge & Kegan Paul.

Carothers, Courtney. 2008. "'Rationalized Out': Discourses and Realities of Fisheries Privatization in Kodiak, Alaska." Pp. 55–74 in *Enclosing the Fisheries: People, Places, and Power: American Fisheries Society Symposium 68,* ed. M. Lowe and C. Carothers. Bethesda, MD: American Fisheries Society.

Carothers, Courtney. 2010. "Tragedies of Commodification: Transitions in Alutiiq Fishing Communities in the Gulf of Alaska." *MAST* 9, no. 2: 95–120.

Carothers, Courtney. 2011. "Equity and Access to Fishing Rights: Exploring the Community Quota Program in the Gulf of Alaska." *Human Organization* 70, no. 3: 213–223.

Carothers, Courtney. 2012. "Enduring Ties: Salmon and the Alutiiq/Sugpiaq Peoples of the Kodiak Archipelago, Alaska." Pp. 133–160 in *Keystone Nations: Indigenous Peoples and Salmon across the North Pacific,* ed. B. J. Colombi and J. F. Brooks. Santa Fe, NM: School for Advanced Research Press.

Carothers, Courtney. In press. A survey of US halibut IFQ holders: Market participation, attitudes, and impacts. Marine Policy (2012), http://dx.doi.org/10.1016/j.marpol.2012.08.007.

Carothers, Courtney, Daniel K. Lew, and Jennifer Sepez. 2010. "Fishing Rights and Small Communities: Alaska Halibut IFQ Transfer Patterns." *Ocean & Coastal Management* 53, no. 9: 518–523.

Castree, Noel. 2010. "Neoliberalism and the Biophysical Environment: A Synthesis and Evaluation of the Research." *Environment and Society: Advances in Research* 1: 5–45.

Christy, Francis. 1973. "Fishermen Quotas: A Tentative Suggestion for Domestic Management." Law of the Sea Institute Occasional Paper 19, University of Rhode Island, Kingston.

Chu, Cindy. 2009. "Thirty Years Later: The Global Growth of ITQs and Their Influence on Stock Status in Marine Fisheries." *Fish and Fisheries* 10, no. 2: 217–230.

Ciriacy-Wantrup, Sigfried von, and Richard C. Bishop. 1975. "'Common Property' as a Concept in Natural Resources Policy." *Natural Resources Journal* 15, no. 4: 713–727.

Community Fisheries Network. 2012. "Community Fisheries Network." http://www.communityfisher iesnetwork.org (accessed 18 May 2012).

Connor, Robin, and Dave Alden. 2001. "Indicators of the Effectiveness of Quota Markets: The South East Trawl Fishery of Australia." *Marine Freshwater Research* 52, no. 4: 387–397.

Copes, Parzival. 1986. "A Critical Review of the Individual Quota as a Device in Fisheries Management." *Land Economics* 62, no. 3: 278–291.

Copes, Parzival. 2000. "Adverse Impacts of Individual Quota Systems on Conservation and Fish Harvest Productivity." Simon Fraser University, Institute of Fisheries Analysis, Discussion Paper 00-2. http://oregonstate.edu/dept/IIFET/copes_morocco.pdf (accessed 8 January 2012).

Copes, Parzival, and Anthony Charles. 2004. "Socioeconomics of Individual Transferable Quotas and Community-Based Fishery Management." *Agricultural and Resource Economics Review* 33, no. 2: 171–181.

Copes, Parzival, and Gísli Pálsson. 2000. "Challenging ITQs: Legal and Political Action in Iceland, Canada and Latin America: A Preliminary Overview." *IIFET 2000 Proceedings:* 1–6.

Costello, Christopher, Steven D. Gaines, and John Lynham. 2008. "Can Catch Shares Prevent Fisheries Collapse?" *Science* 321, no. 5896: 1678–1681.

Criddle, Keith, and Seth Macinko. 2000. "A Requiem for the IFQ in US Fisheries?" *Marine Policy* 24, no. 6: 461–469.

Crutchfield, James A. 1979. "Economic and Social Implications of the Main Policy Alternatives for Controlling Fishing Effort." *Journal of the Fisheries Research Board of Canada* 36, no. 7: 742–752.

Davis, Anthony. 1996. "Barbed Wire and Bandwagons: A Comment on ITQ Fisheries Management." *Reviews in Fish Biology and Fisheries* 6, no. 1: 97–107.

Davis, Anthony, and Svein Jentoft. 2001. "The Challenge and the Promise of Indigenous Peoples' Fishing Rights—From Dependency to Agency." *Marine Policy* 25, no. 3: 223–237.

Dewees, Christopher M. 1997. "New Zealand Fishing Industry Changes for 'Pakeha' and Maori with Individual Transferable Quotas." Pp. 91–106 in *Social Implications of Quota Systems in Fisheries,* ed. G. Pálsson and G. Pétursdóttir. Copenhagen: Nordic Council of Ministers.

Dewees, Christopher M. 2008. "Attitudes, Perceptions, and Adaptations of New Zealand Commercial Fishermen During 20 Years of Individual Transferable Quotas." Pp. 35–54 in *Enclosing the Fisheries: People, Places, and Power Symposium 68,* ed. M. Lowe and C. Carothers. Bethesda, MD: American Fisheries Society.

Ecotrust. 2011. "Community Dimensions of Fisheries Catch Share Programs: Integrating Economy, Equity, and Environment." National Panel on the Community Dimensions of Fisheries Catch Share Programs. http://www.ecotrust.org/fisheries/NPCDFCSP_paper_031511.pdf (accessed 18 May 2012).

Einarsson, Níels. 2009. "From Good to Eat to Good to Watch: Whale Watching, Adaptation and Change in Icelandic Fishing Communities." *Polar Research* 28, no. 1: 129–138.

Einarsson, Níels. 2011. "Fisheries Governance and Social Discourse in Post-Crisis Iceland: Responses to the UN Human Rights Committee's Views in case 1306/2004." Pp. 107–142 in *Culture, Conflict and Crises in the Icelandic Fisheries. An Anthropological Study of People, Policy and Marine Resources in the North Atlantic Arctic.* Uppsala Studies in Cultural Anthropology 48. Uppsala, Sweden: Acta Universitatis Upsaliensis.

Environmental Defense Fund (EDF). 2012. "Catch Shares—New Hope for Fisheries." http://www.edf.org/oceans/catch-shares (accessed 6 January 2012).

Eythórsson, Einar. 1996. "Theory and Practice of ITQs in Iceland Privatization of Common Fishing Rights." *Marine Policy* 20, no. 3: 269–281.

Food and Agriculture Organization (FAO). 2010. *The State of World Fisheries and Aquaculture.* Rome.

Festa, David, Diane Regas, and Judson Boomhower. 2008. "Sharing the Catch, Conserving the Fish." *Issues in Science and Technology* (Winter): 75–84.

Fina, Mark. 2011. "Evolution of Catch Share Management: Lessons from Catch Share Management in the North Pacific." *Fisheries* 36, no. 4: 164–177.

Fujita, Rod M., and Kate Bonzon. 2006. "Rights-Based Fisheries Management: An Environmentalist's Perspective." *Reviews in Fish Biology and Fisheries* 15, no. 3: 309–312.

Gordon, H. Scott. 1954. "The Economic Theory of a Common Property Resource: The Fishery." *Journal of Political Economy* 62, no. 2: 124–142.

Hannesson, Rögnvaldur. 2005. "Rights Based Fishing: Use Rights Versus Property Rights to Fish." *Reviews in Fish Biology and Fisheries* 15, no. 3: 231–241.

Hannesson, Rögnvaldur. 2006. *The Privatization of the Oceans.* Cambridge: MIT Press.

Hardin, Garret. 1968. "The Tragedy of the Commons." *Science* 162, no. 3859: 1243–1248.

Hartley, Marcus, and Mark Fina. 2001. "Changes in Fleet Capacity Following the Introduction of Individual Vessel Quotas in the Alaskan Pacific Halibut and Sablefish Fishery." Pp. 186–207 in *Case Studies on the Allocation of Transferable Quota Rights in Fisheries, FAO Technical Paper No. 411,* ed. R. Shotton. Rome: FAO.

Harvey, David. 2003. *The New Imperialism.* New York: Oxford University Press.

Heal, Geoffrey, and Wolfram Schlenker. 2008. "Sustainable Fisheries." *Nature* 455, no. 7216: 1044–1045.

Helgason, Agnar, and Gísli Pálsson. 1997. "Contested Commodities: The Moral Landscape of the Modernist Regime." *Journal of the Royal Anthropological Institute* 3, no. 3: 451–471.

Helgason, Agnar, and Gísli Pálsson. 1998. "Cash for Quotas: Disputes over the Legitimacy of an Economic Model of Fishing in Iceland." Pp. 117–134 in *Virtualism: A New Political Economy,* ed. J. Carrier and D. Miller. New York: Berg.

Herrmann, Mark, and Keith Criddle. 2006. "An Econometric Market Model for the Pacific Halibut Fishery." *Marine Resource Economics* 21, no. 2: 129–158.

Heynen, Nik, James McCarthy, Scott Prudham, and Paul Robbins. 2007. "Introduction: False Promises." Pp. 1–21 in *Neoliberal Environments: False Promises and Unnatural Consequences,* ed. N. Heynen, J. McCarthy, S. Prudham, and P. Robbins. New York: Routledge.

Holm, Petter, and Kåre N. Nielsen. 2007. "Framing Fish, Making Markets: The Construction of Individual Transferable Quotas (ITQs)." *Sociological Review* 55, no. 2: 173–195.

Icelandic Ministry of Fisheries. 2011. "Fisheries Information: Quotas." http://www.fisheries.is/management/fisheries-management/individual-transferable-quotas/ (accessed 5 January 2012).

Jentoft, Svein, and G. Karlsen. 1996. "Sami Fisheries, Quota Management, and the Rights Issues." Pp. 147–164 in *Social Implications of Quota Systems in Fisheries,* ed G. Pálsson and G. Pétursdóttir. Copenhagen: Nordic Council of Ministers.

Karlsdóttir, Anna. 2008. "Not Sure about the Shore! Transformation Effects of Individual Transferable Quotas on Iceland's Fishing Economics and Communities." Pp. 99–117 in *Enclosing the Fisheries: People, Places, and Power: American Fisheries Society Symposium 68,* ed. M. Lowe and C. Carothers. Bethesda, MD: American Fisheries Society.

Knapp, Gunnar. 2011. "Local Permit Ownership in Alaska Salmon Fisheries." *Marine Policy* 35, no. 5: 658–666.

Langdon, Stephen J. 2008. "The Community Quota Program in the Gulf of Alaska: A Vehicle for Alaska Native Village Sustainability?" Pp. 155–194 in *Enclosing the Fisheries: People, Places, and Power: American Fisheries Society Symposium 68,* ed. M. Lowe and C. Carothers. Bethesda, MD: American Fisheries Society.

Latour, Bruno. 1987. *Science in Action: How to Follow Scientists and Engineers Through Society.* Cambridge, MA: Harvard University Press.

Lazrus, Heather M., Jennifer A. Sepez, Ron G. Felthoven, and J. C. Lee. 2011. "Post-Rationalization Restructuring of Commercial Crew Member Opportunities in Bering Sea and Aleutian Island Crab Fisheries." NOAA Technical Memorandum NMFS-AFSC-217. http://www.afsc.noaa.gov/Publications/AFSC-TM/NOAA-TM-AFSC-217.pdf (accessed 6 January 2012).

Lowe, Marie E., and Courtney Carothers, eds. 2008. *Enclosing the Fisheries: People, Places, and Power: American Fisheries Society Symposium 68.* Bethesda, MD: American Fisheries Society.

Macinko, Seth, and Daniel Bromley. 2002. *Who Owns America's Fisheries?* Washington, DC: Island Press.

Macinko, Seth, and Daniel Bromley. 2004. "Property and Fisheries for the Twenty-First Century: Seeking Coherence from Legal and Economic Doctrine." *Vermont Law Review* 28, no. 3: 623–661.

Macinko, Seth, and Sarah Schumann. 2008. "The Process of 'Property': Stasis and Change in Lobster Management in Southern New England." *Vermont Law Review* 33, no. 1: 73–129.

Mansfield, Becky. 2004. "Neoliberalism in the Oceans: 'Rationalization,' Property Rights, and the Commons Question." *Geoforum* 35, no. 3: 313–326.

Mansfield, Becky. 2008. "Introduction: Property and the Remaking of Nature-Society Relations." Pp. 1–13 in *Privatization: Property and the Remaking of Nature-Society Relations,* ed. B. Mansfield. Malden, MA: Blackwell.

Mansfield, Becky. 2010. "'Modern' Industrial Fisheries and the Crisis of Overfishing." Pp. 84–99 in *Global Political Ecology,* ed. R. Peet, P. Robbins and M. Watts. New York: Routledge.

Maurstad, Anita. 2000. "To Fish or Not to Fish: Small-Scale Fishing and Changing Regulations of the Cod Fishery in Northern Norway." *Human Organization* 59, no. 1: 37–47.

McCarthy, J., and S. Prudham. 2004. "Neoliberal Nature and the Nature of Neoliberalism." *Geoforum* 35, no. 3: 275–283.

McCay, Bonnie J. 1995. "Social and Ecological Implications of ITQs: An Overview." *Ocean and Coastal Management* 28, no. 13: 3–22.

McCay, Bonnie J. 2001. "Initial Allocation of Individual Transferable Quotas in the US Surf Clam and Ocean Quahog Fishery." Pp. 86–90 in *Case Studies on the Allocation of Transferable Quota Rights in Fisheries, FAO Technical Paper No. 411,* ed. R. Shotton. Rome: FAO.

McCay, Bonnie J. 2004. "ITQs and Community: an Essay on Environmental Governance." *Agricultural and Resource Economics Review* 33, no. 2: 162–170.

McCay, Bonnie J., and James M. Acheson, eds. 1987. *The Question of the Commons: The Culture and Ecology of Communal Resources.* Tucson: University of Arizona Press.

McCay, Bonnie, Carolyn Creed, Alan Finlayson, Richard Apostle, and Knut Mikalsen. 1995. "Individual Transferable Quotas (ITQs) in Canadian and US Fisheries." *Ocean and Coastal Management* 28, no. 1-3: 85–115.

McCormack, Fiona. 2007. "Moral Economy and Maori Fisheries." *SITES: New Series* 4, no. 1: 45–69.

McCormack, Fiona. 2010. "Fish is My Daily Bread: Owning and Transacting in Maori Fisheries." *Anthropological Forum* 20, no. 1: 19–39.

McEvoy, Arthur F. 1986. *The Fishermen's Problem: Ecology and Law in the California Fisheries, 1850–1980.* Cambridge: Cambridge University Press.

McGoodwin, James R. 1990. *Crisis in the World's Fisheries: People, Problems, and Policies.* Stanford, CA: Stanford University Press.

Melnychuk, Michael C., Timothy E. Essington, Trevor A. Branch, Selina S. Heppell, Olaf P. Jensen, Jason S. Link, Steven J. D. Martell, Ana M. Parma, John G. Pope, and Anthony D. M. Smith. 2011. "Can Catch Share Fisheries Better Track Management Targets?" *Fish and Fisheries* doi: 10.1111/j.1467-2979.2011.00429.x.

Memon, P. Ali, and Ross Cullen. 1992. "Fishery Policies and Their Impact on the New Zealand Maori." *Marine Resource Economics* 7, no. 3: 153–167.

Memon, P. Ali, and Nicholas A. Kirk. 2011. "Maori Commercial Fisheries Governance in Aotearoa/New Zealand within the Bounds of a Neoliberal Fisheries Management Regime." *Asia Pacific Viewpoint* 52, no. 1: 106–118.

Moloney, David G., and Peter H. Pearse. 1979. "Quantitative Rights as an Instrument for Regulating Commercial Fisheries." *Journal of the Fisheries Research Board of Canada* 36, no. 7: 859–866.

National Research Council (NRC). 1999a. *Sharing the Fish: Towards a National Policy on Individual Fishing Quotas.* Washington, DC: National Academy Press.

National Research Council. 1999b. *The Community Development Quota Program in Alaska.* Washington, DC: National Academy Press.

NOAA Fisheries Service. 1993. *Pacific Halibut Fisheries; Groundfish of the Gulf of Alaska; Groundfish of the Bering Sea and Aleutian Islands, Limited Access Management of Fisheries Off Alaska.* Final Rule. 58 Federal Register 215.

NOAA Fisheries Service. 2010. *Report on Holdings of Individual Fishing Quota (IFQ) by Residents of Selected Gulf of Alaska Fishing Communities.* Juneau, AK: NOAA Fisheries Service.

NOAA Fisheries Service. 2012. "NOAA Knows … Catch Share Programs." http://www.noaa.gov/factsheets/new%20version/catch_shares.pdf (accessed 18 January 2012).

Olson, Julia. 1997. "The Cultural Politics of Fishing: Negotiating Community and Common Property in Northern Norway." PhD diss., Stanford University.

Olson, Julia. 2011. "Understanding and Contextualizing Social Impacts from the Privatization of Fisheries: An Overview." *Ocean & Coastal Management* 54, no. 5: 353–363.

Ostrom, Elinor, Thomas Dietz, Nives Dolsak, Paul C. Stern, Susan Stonich, and Elke U. Weber, eds. 2002. *The Drama of the Commons: Committee on the Human Dimensions of Global Change.* Washington, DC: National Academy Press.

Pálsson, Gísli, and Agnar Helgason. 1995. "Figuring Fish and Measuring Men: The ITQ System in the Icelandic Cod Fishery." *Ocean & Coastal Management* 28, no. 1-3: 117–146.

Pálsson, Gísli, and Agnar Helgason. 1996. "The Politics of Production: Enclosure, Equity and Efficiency." Pp. 60–86 in *Images of Contemporary Iceland,* ed. G. Pálsson and P. Durrenburger. Iowa City: University of Iowa Press.

Pálsson, Gísli, and Gudrún Pétursdóttir, eds. 1996. *Social Implications of Quota Systems in Fisheries.* Copenhagen: Nordic Council of Ministers.

Pautkze, Clarence, and Chris Oliver. 1997. *Development of the Individual Fishing Quota Program for Sablefish and Halibut Longline Fisheries off Alaska.* Anchorage, AK: North Pacific Fishery Management Council.

Pinkerton, Evelyn, and Danielle Edwards. 2009. "The Elephant in the Room: The Hidden Costs of Leasing Individual Transferable Fishing Quotas." *Marine Policy* 33, no. 4: 707–713.

Polanyi, Karl. 1944. *The Great Transformation: The Political and Economic Origins of Our Time.* Boston: Beacon Press.

Reedy-Maschner, Katherine. 2010. *Aleut Identities: Tradition and Modernity in an Indigenous Fishery.* Montreal: McGill-Queen's University Press.

Reiser, Alison. 1999. "Prescriptions for the Commons: Environmental Scholarship and the Fishing Quotas Debate." *Harvard Environmental Law Review* 23, no. 2: 393–420.

Rosvold, Eric. 2007. "Graying of the Fleet: Community Impacts of Asset Transfers." Pp. 67–72 in *Alaska's Fishing Communities: Harvesting the Future,* ed. P. Cullenburg. Fairbanks: Alaska Sea Grant College Program, University of Alaska Fairbanks.

Shotton, Ross, ed. 2000a. "Use of Property Rights in Fisheries Management." Proceedings of the FishRights99 Conference, Fremantle, Western Australia, 11–19 November 1999. Mini-course and Core Conference Presentations.FAO Fisheries Technical Paper 404/1, Rome.

Shotton, Ross, ed. 2000b. "Use of Property Rights in Fisheries Management." Proceedings of the FishRights99 Conference, Fremantle, Western Australia, 11–19 November 1999. Workshop Papers. FAO Fisheries Technical Paper 404/2, Rome.

Shotton, Ross, ed. 2001. "Case Studies on the Allocation of Transferable Quota Rights in Fisheries." FAO Technical Paper No. 411, Rome.

Skaptadóttir, Unnur D. 2000. "Women Coping With Change in an Icelandic Fishing Community: A Case Study." *Women's Studies International Forum* 23, no. 3: 311–321.

St. Martin, Kevin. 2005. "Disrupting Enclosure in New England Fisheries." *Capitalism, Nature, Socialism* 16, no. 1: 63–80.

St. Martin, Kevin. 2007a. "The Difference That Class Makes: Neoliberalization and Noncapitalism in the Fishing Industry of New England." *Antipode* 39, no. 3: 527–549.

St. Martin, Kevin. 2007b. "Enclosure and Economic Identity in New England Fisheries." Pp. 255–266 in *Neoliberal Environments: False Promises and Unnatural Consequences,* ed. N. Heynen, J. McCarthy, S. Prudham, and P. Robbins. New York: Routledge.

St. Martin, Kevin. 2008. "The Difference that Class Makes: Neoliberalization and Non-Capitalism in the Fishing Industry of New England." Pp. 133–155 in *Privatization: Property and the Remaking of Nature-Society Relations,* ed. B. Mansfield. Malden, MA: Blackwell.

Stewart, James, and Peter Callagher. 2011. "Quota Concentration in the New Zealand Fishery: Annual Catch Entitlement and the Small Fisher." *Marine Policy* 3, no. 5: 631–646.

Stewart, James, Kim Walshe, and Beverly Moodie. 2006. "The Demise of the Small Fisher? A Profile of Exiters from the New Zealand Fishery." *Marine Policy* 30, no. 4: 328–340.

Weiss, Kenneth. 2008. "Sharing the Catch Helps Save the Fish." *Los Angeles Times,* 19 September. http://articles.latimes.com/2008/sep/19/nation/na-fish19.

Wilen, James E. 2000. "Renewable Resource Economists and Policy: What Differences Have We Made?" *Journal of Environmental Economics and Management* 39, no. 3: 306–327.

Wilk, Richard. 1996. *Economies and Cultures: Foundations of Economic Anthropology.* Boulder, CO: Westview Press.

Wingard, John D. 2000. "Finding Alternatives to Privatizing the Resource: Community-Focused Management for Marine Fisheries." Pp. 5–29 in *Communities and Capital: Local Struggles Against Corporate Power and Privatization,* ed. T. Collins and J. Wingard. Athens: University of Georgia Press.

Yandle, Tracy, and Christopher M. Dewees. 2008. "Consolidation in an Individual Transferable Quota Regime: Lessons from New Zealand, 1986–1999." *Environmental Management* 41, no. 6: 915–928.

Contradictions in Tourism

The Promise and Pitfalls of Ecotourism as a Manifold Capitalist Fix

Robert Fletcher and Katja Neves

ABSTRACT: This article reviews an interdisciplinary literature exploring the relationship between tourism and capitalism focused on ecotourism in particular. One of this literature's most salient features is to highlight ecotourism's function in employing capitalist mechanisms to address problems of capitalist development itself by attempting to resolve a series of contradictions intrinsic to the accumulation process, including: economic stagnation due to overaccumulation (time/space fix); growing inequality and social unrest (social fix); limitations on capital accumulation resulting from ecological degradation (environmental fix); a widespread sense of alienation between humans and nonhuman natures; and a loss of "enchantment" due to capitalist rationalization. Hence, widespread advocacy of ecotourism as a "panacea" for diverse social and environmental ills can be interpreted as an implicit endorsement of its potential as a manifold capitalist fix as well. The article concludes by outlining a number of possible directions for future research suggested by this review.

KEYWORDS: accumulation; body; capitalism; desire; ecotourism; neoliberalism

I stand on the pier waiting for yet another group of tourists to arrive and be loaded onto the 12-person zodiac. This will be the Moby Dick's fourth voyage of the day, its two hundred twenty-fourth of the season—and we are only two months in. It is June and if all goes well the season should last until early October. The tourists come walking down the pier, already wearing their inflated life vests—and over them their video and photo cameras, hats, sweaters, backpacks, and raincoats, despite the fact that it is a perfectly sunny day. They are eager to come onboard though I can see from their dangerous bouncing on the zodiac's bow that they are not used to being at sea. After a considerable amount of disorderly excitement they finally sit down and we leave port.

That day it takes about two hours to find the whales. Far more than usual, and to make matters worse, there are no dolphins to be found on the way to distract and amuse our tourists. This is too bad. Some of the tourists are on gravol (seasickness medication) and by the time we reach the whales they are a little slow in reacting. The crew came prepared. A bit of coffee soon has everyone back on track. The skipper Rita (a former whale hunter) and his first mate Louisa (a marine biologist) tell the tourists where to look. In the middle of the ocean it is not that easy to spot a sperm whale. When the sun is this bright and the air is misty all one can see is a dim whale spout. Other than that, one sees only a bit of a mass almost the same color as the water. Whale watching rules created to protect the cetaceans do not allow the boat to get too close, otherwise one might get a better glimpse; the whale may choose to come closer of course, but that rarely happens.

Environment and Society: Advances in Research 3 (2012): 60–77 © Berghahn Books
doi:10.3167/ares.2012.030105

The tourists wait, not sure what to see, or how to enjoy what to them looks like a nondescript floating blob. They tell me "It's not quite what I expected" … "I can hardly tell it's a whale" … The skipper and his mate try to explain what is so interesting about the whales and what they are doing, and how the passengers might enjoy just sitting there observing them. They tell everyone they should be excited! Thrilled! Happy! They do it passionately, even though this is their fourth voyage of the day, their two hundred twenty-fourth of the season … Their efforts fall on mute ears. Even though these tourists are on their first trip, the marketing ads, internet, word of mouth, and so on have convinced them that the truly cathartic and spiritually meaningful moment of this trip will happen when the whales fluke while diving deep into the ocean. They are eager to experience that moment. They are also getting tired, and a little seasick. The skipper gives in and even though he is not supposed to, he gently navigates the boat such as to "guide the whales into fluking." As if led by an invisible orchestrator the cameras flash in unison for a few moments until the last bit of fluke submerges. Every single tourist displays a smile of satisfaction and victory. They are all done.

The preceding vignette, from Neves's long-term ethnographic research on cetacean tourism (e.g., Neves 2004, 2006, 2010) captures one of the main dynamics we address in this article: ecotourism's capacity to transform bodies into sites of virtually limitless capital accumulation by promising a satisfying experience yet usually delivering instead a mere "pseudocatharsis" (Neves 2009a) that paradoxically stimulates a desire for further experience in pursuit of the fulfillment continually deferred. Documentation of this dynamic adds a new dimension to previous research analyzing ecotourism's impressive capacity to provide a "fix" of sorts for a variety of contradictions inherent in the accumulation process. In this state-of-the-field review, we describe this and other research investigating the multidimensional relationship between ecotourism and capitalism, in keeping with this journal issue's special focus on capitalism and the environment. Rather than provide an exhaustive review of the ecotourism literature as a whole, we focus on the research most pertinent to this specific theme.

After surveying the current literature exploring the ecotourism-capitalism relationship we push the analysis further by drawing on our own ongoing ethnographic fieldwork to describe several more aspects yet to be addressed by this research, one of the most intriguing of which involves treating the human body as a prime site of generative accumulation through commodification of a particular affective state. We conclude by highlighting the implications of our analysis for further research concerning the conjunction of ecotourism and capitalism, outlining several productive avenues that might be pursued in future study.

Explaining Ecotourism

The ecotourism industry has been growing rapidly over the past several decades. By 1998 the United Nations World Tourism Organization (UNWTO) estimated that ecotourism comprised 20 percent of the US$441 billion global tourism market and was growing approximately 30 percent per year (versus 4% for the industry as a whole) (UNWTO 1998). In 2004, the UNWTO reported again that ecotourism was continuing to develop at three times the industry average (The International Ecotourism Society 2006). In short, Honey observes, "Ecotourism is often claimed to be the most rapidly expanding sector of the tourism industry" (2008: 6).

Of course, how the industry is measured depends on how "ecotourism" is defined. In its popular usage, the term is virtually synonymous with nature-based tourism broadly conceived, and thus "covers many activities: visiting a national park in Montana, diving in the Caribbean, seeing Mayan ruins, staying at a village lodge in Papua New Guinea" (West and Carrier 2004:

491). Yet a growing movement seeks to conceptualize ecotourism more narrowly as only nature-based tourism that confers significant social and environmental benefits (see esp. Honey 2008). Hence, The International Ecotourism Society, in a widely cited definition, characterizes ecotourism as "Responsible travel to natural areas that conserves the environment and improves the well-being of local people" (cited in Honey 2008: 6).

Our aim here is not to endorse either position but rather to analyze the type of activities that comprise ecotourism's common quest for encounters with nonhuman natures;[1] hence we define the phenomenon broadly for purposes of this discussion. Regardless of one's preferred definition, it is clear that ecotourism has become a global industry of significant proportions. As a result, it has become a popular strategy for sustainable development and environmental conservation around the world, enthusiastically promoted by international financial institutions (IFIs), national governments, nongovernmental organizations (NGOs), academic researchers, industry professionals, and innumerable local community members alike (Mowforth and Munt 2008). Honey summarizes:

> Around the world, ecotourism has been hailed as a panacea: a way to fund conservation and scientific research, protect fragile and pristine ecosystems, benefit rural communities, promote development in poor countries, enhance ecological and cultural sensitivity, instill environmental awareness and a social conscience in the travel industry, satisfy and educate the discriminating tourist, and, some claim, build world peace. (2008: 4)

Further emphasizing this potential, the United Nations declared 2002 the International Year of Ecotourism (see Butcher 2006b), highlighting the "the need for international cooperation in promoting tourism within the framework of sustainable development."[2]

Several explanations have been offered to account for ecotourism's dramatic surge in growth and popularity over the past several decades (see Fletcher 2011). From the demand side, researchers point to the behavior of "new" or "alternative" tourists (Mowforth and Munt 2008; Poon 1993) from wealthy industrial societies who, since the 1970s, have become increasingly "[t]urned off by overcrowded, unpleasant conditions" at home and have thus began seeking "serenity and pristine beauty overseas" (Honey 2008: 12; see also Fletcher 2009; Mowforth and Munt 2008). On the supply side, ecotourism is widely considered a superior alternative to conventional mass tourism, ostensibly compensating for the numerous negative impacts (including increased crime, drug use, prostitution, leakage—flow of most revenue out of the local area—and environmental degradation) conventional tourism development commonly entails (Honey 2008; Mowforth and Munt 2008). Ecotourism is considered particularly conducive to small-scale sustainable development in rural areas of less developed societies because it generally targets precisely these areas, which will lose their competitive advantage if overdevelopment occurs (West and Carrier 2004). In this spirit, ecotourism is championed as a model sustainable development strategy for areas that have not experienced significant benefits from conventional development measures, leading Munt (1994: 49) to describe it "as a last-ditch attempt to break from the confines of underdevelopment and get the IMF to lay the golden egg of an upwardly-mobile GNP."

In this article, we explore a third, complementary line of analysis that attributes both demand- and supply-side aspects of ecotourism's growth in popularity to its function as a particular form of capitalism that offers "fixes" for a series of contradictions inherent to the process of capitalist accumulation. In this sense, the many dimensions of the lofty promise commonly attributed to ecotourism, as highlighted by Honey (2008), can be interpreted as implicitly referencing the manifold fix that ecotourism promises for the capitalist world economy. We contend, therefore, that in the contemporary era ecotourism development has become an important means by which capitalism endeavors to overcome its own contradictions (see Duffy 2012; Fletcher 2011). We also argue that this process is itself contradictory because ecotourism attempts to overcome these contradictions by using the very mechanisms and capitalist processes that created them. Paradoxically, therefore, not only does ecotourism development reproduce fundamental contradictions of capitalist accumulation, it also generates a series of further contradictions. We describe these, observing the common construction of an ecotourism "bubble" or "script" intended to conceal inconsistencies from potential clients and funders in order to preserve an image of success and keep the finances flowing.

Although our ethnographic research focuses on particular, fairly dramatic forms of ecotourism (whale watching and whitewater rafting), we believe our analysis applies to a wide range of nature-based tourism pursuits. Many of the dynamics we describe can be found in more conventional, mass forms of tourism as well (see Fletcher 2011). Yet given this special issue's focus on capitalism and the environment we limit our analysis to ecotourism specifically.

Ecotourism as an Accumulation Strategy

The tourism industry as a whole has long been described as "a product of metropolitan capitalist enterprise" (Britton 1982: 331) and "a major internationalized component of Western capitalist economies" (Britton 1991: 451). Ecotourism in particular is often considered the cutting edge of this trend, facilitating the increased commodification of natural resources around the globe (Bandy 1996; West and Carrier 2004). In this analysis, ecotourism is commonly categorized as part of a "third wave" of tourism development as the industry has evolved in concert with global capitalism (Lash and Urry 1987; Urry 2001). In its origins as a small-scale, elite enterprise, tourism of the nineteenth-century Grand Tour variety reflected early liberal capitalism's nascent entrepreneurial structure. The rise of mass tourism centered on collective prepackaged holidays in the post–World War II era, by contrast, coincided with the consolidation of an "organized," Fordist regime of accumulation emphasizing increasingly larger, vertically integrated firms. Finally, the 1970s witnessed the rise of "new" or "alternative" tourism offering a variety of flexible, individually tailored trips concurrent with capitalism's shift toward a novel "disorganized," post-Fordist form centered on "flexible accumulation" (Harvey 1989) through diverse structures. This has led to the development of a myriad "niche" or "boutique" markets designed to offer an outlet for every tourist's particular taste, including such diverse (and disturbing) products as war, sex, and slum tourism (Gibson 2009; Munt 1994).

In this sense, ecotourism is implicated in the emergence of what Martin O'Connor (1994) calls capitalism's "ecological phase" transitioning from the "formal" to "real" subsumption of nature within production (Smith 2007). This new ecological phase, of course, is itself part and parcel of capitalism's neoliberal turn since the 1970s (Brockington et al. 2008). Ecotourism, therefore, has been described as an expression of neoliberalization as well, embodying such paradigmatic free market principles as decentralization and deregulation of natural resource governance (or, more precisely, *re*regulation from states to nonstate actors) as well as resources' marketization, privatization, and commodification as tourism "products" (see Bianchi 2005, 2009; Carrier and Macleod 2005; Cater 2006; Davis 1997; Duffy 2002, 2008, 2010, 2012; Duffy and Moore 2010; Fletcher 2009, 2011; Mowforth and Munt 2008; Neves 2010; Vivanco 2001, 2006; West and Carrier 2004). West and Carrier (2004: 484) thus characterize ecotourism as "the institutional expression of particular sets of late capitalist values in a particular political-economic climate," while Cater (2006) similarly labels ecotourism a "Western construct" expanding the hegemony of global capitalism. Duffy (2012: 17) goes further to contend that ecotourism "is not just reflective of global neoliberalism, but constitutes one of its key drivers, extending neoliberal principles to an expanding range of biophysical phenomena."

This analysis is part of a growing body of research describing an increasing trend toward neoliberalization within natural resource management in general around the world. Initially this research centered on conventional forms of resource extraction and processing (e.g., Bakker 2009; Castree 2008, 2010; Heynen et al. 2007; McCarthy and Prudham 2004). However, more recently it has turned its focus to environmental conservation in particular (e.g., Brockington and Duffy 2010; Brockington et al. 2008; Büscher 2010; Büscher et al. 2012; Dressler and Roth 2010; Fletcher 2010b; Igoe and Brockington 2007; Neves 2010; Sullivan 2006, 2009). While extractive industry creates value by transforming natural resources into commodities that can be transported to their point of consumption, conservation, by contrast, seeks to commodify resources in situ, necessitating particular mechanisms to generate value sans extraction (Büscher et al. 2012). By transporting consumers to the point of production where they pay to interact with preserved resources, ecotourism thus serves as an—currently perhaps the most—important financing mechanism for neoliberal conservation.

Ecotourism's promotion as a conservation strategy is often based in an explicitly neoliberal approach to human governance in general, which asserts that if sufficient economic value is attached to in situ resources local stakeholders will be incentivized to preserve rather than extract them (see Fletcher 2010b). Honey (2008: 14) calls this the "stakeholder theory" asserting "that people will protect what they receive value from." This perspective is repeated ad nauseum in both academic literature and popular press (Fletcher 2009; Stronza 2007). As but one example, Crapper (1998: 21) contends of an ecotourism project in Peru, "As more native communities start to reap direct economic benefits as owners and partners of tourism services, locals will have more of an incentive, and a challenge, to protect what the tourists come to see." In reality, however, researchers have shown that such benefits are usually spread unevenly, often deepening preexisting social inequalities, and even introducing serious problems of equity and social justice (Brockington et al. 2008; Fletcher 2012; Neves and Igoe 2012; Stronza 2007).

The tourism industry as a whole has been described as an important means by which the capitalist world economy has sought to sustain itself through geographic and temporal expansion in the postwar era (Fletcher 2011); and as an expression of neoliberal capitalism, ecotourism in particular is seen to offer a number of potential "fixes" (Harvey 1989, 2006) addressing contradictions inherent to the accumulation process (Cater 2006; Fletcher 2011). As Marx (1973) observed, at the heart of the capitalist economy stands a central contradiction between competing motives of production and consumption. Capitalists' aim to extract maximum profit from the production process periodically precipitates a crisis of "overaccumulation" or "overproduction" in which workers, in aggregate, lack funds to absorb the fruits of production, causing profits to fall and production to stagnate. To transcend this crisis in the short term, excess accumulated capital must be reinvested in profitable production, through geographic expansion (what Harvey calls a "spatial fix"); through a "temporal fix" entailing either investment with the promise of future return or reducing turnover time so that "speed-up this year absorbs excess capacity from last year" (Harvey 1989: 182); or through a combination of these (a "time-space fix") primarily involving international money lending.

This expansion in response to Marx's central contradiction precipitates another crisis, which James O'Connor (1988, 1994) calls capitalism's "second contradiction" following from the reality that production is ultimately predicated on a finite natural resource base. Eventually, increased production in order to recover profit in the face of an overaccumulation crisis will degrade this resource base, causing costs to rise and profits to fall once more. This second contradiction can be addressed—and profit temporarily restored—through what Castree (2008) calls a series of "environmental fixes," involving commodification and trading new forms of "natural capital"; replacing state control of resources with capitalist markets; intensifying exploitation of a given natural resource to yield increased short-term profits; or transferring resource governance responsibility (and thus revenues) from states to nonstate actors.

Ecotourism offers opportunities in relation to all these fixes (Fletcher 2011).[3] Development of new ecotourism destinations delivers a spatial fix, while investment in new ventures provides a temporal fix. By selling a transient event that is instantly consumed, ecotourism offers an additional temporal fix through reducing turnover time for the recovery of invested capital to a minimum. International lending for ecotourism development, such as provided by the World Bank, presents a time-space fix as well. In its status as a service industry, ecotourism offers a further opportunity for addressing overaccumulation crises (Fletcher 2011). Service work separates producers from consumers, allowing capitalists to extract maximum profit from the production process without compromising the consumer base necessary to forestall crisis by facilitating the transfer of a portion of accumulated capital to service workers—who are then able to absorb production—in exchange for the latter's provision of an additional valued benefit.

Ecotourism in particular offers a number of potential environmental fixes (Fletcher 2011; Robbins and Fraser 2003). It creates new markets for natural capital by expanding into the relatively noncommodified spaces that are its ideal destinations. Through privatization (such as in the widespread development of private nature reserves; see Langholz and Lassoie 2001), it transfers resource control from states to capitalist markets. Increasing visitation augments the revenue that can be generated from a given destination and the resources therein. The growing centrality of NGOs and private consultancy firms in the ecotourism development process (as intermediaries, for instance, in the transfer of funds from IFIs to local communities) helps to further shift the locus of revenue generation from states to nonstate players (Butcher 2006a). Ecotourism can be seen to provide a further environmental fix, which neither Castree nor James O'Connor predicted, by actually harnessing resource degradation itself as an additional source of value (Fletcher 2011; Neves 2010)—a process resonant of Klein's (2007) description of neoliberalism in general as a strategy of "disaster capitalism." As resources grow scarce, the remainder become increasingly valuable, and ecotourism destinations are in fact frequently marketed by emphasizing the likelihood that they will cease to exist in the future (Mowforth and Munt 2008). This trend is exacerbated by recent growth in "extinction tourism": the visitation of sites (e.g., glaciers, small island states threatened by sea level rise) whose value derives explicitly from the prospect of their imminent disappearance (see Leahy 2008).

Pushing this line of analysis further, ecotourism development can be seen to provide a variety of other fixes to problems intrinsic to capitalist development. For instance, Doane (2010) describes "fair trade" coffee as offering a "social fix" in its aim to deliver a living wage to producers and thereby redress to a degree the inequality and attendant social unrest created by capitalist markets. Through fair trade, then, inequality is actually harnessed as a source of increased value (in the form of the higher prices that can be charged) resulting from fair trade's very promise to assuage this inequality. In the demand that it redress inequities of uneven development and enhance the well-being of poor, rural community members bypassed by conventional development, ecotourism can be seen to offer a similar social fix.

Ecotourism is commonly marketed specifically as a means to connect humans with nonhuman natures (Braun 2003; Fletcher 2009; West and Carrier 2004), and in this respect it can be understood to offer yet another fix to problems wrought by capitalism (Neves under review). As Marx (1973) observed, capitalist production tends to create a "metabolic rift" whereby both producers and consumers are increasingly divorced from the means of production and the nonhuman natures in which this production is grounded (Bellamy Foster 2000). As a result, capitalist subjects come to experience themselves and the human sphere they occupy as alienated from nonhuman natures altogether. In offering an experience of "nature-culture unity" (Neves under review), ecotourism promises to resolve this division and the alienation it precipitates (Braun 2003), and thus can be described as providing something of a "psychological" fix for this existential crisis created by capitalist development. A further psychological fix can be found in ecotourism's common promise to deliver an extraordinary experience of mystery and enchantment felt to be lacking in everyday life (Arnould and Price 1993; Arnould et al. 1999; Fletcher 2008; Palmer 2004; Stranger 1999). This, too, can be seen as a response to problems of modern capitalist development, which has promoted the progressive disenchantment of the world and establishment of a rational, ordered society devoid of unpredictable elements (Escobar 1995; Tambiah 1990; Weber 1930). Similar extraordinary experiences are of course available through a wide range of activities, from marijuana use (Becker 1953) to spiritual snake handling (Covington 1995), and the problems they address are certainly not limited to the neoliberal age (see Campbell 1987), yet ecotourism in an increasingly prevalent means by which such experiences are pursued in the contemporary period.

In short, ecotourism offers the potential to provide an impressive variety of (partial) resolutions to contradictions of capitalist accumulation, promising spatial, temporal, time-space, environmental, social, and psychological fixes in one concise package. Hence, when Honey (2008) and others describe the multiple dimensions of the potential benefit that ecotourism is commonly called upon to provide, we suggest that this can be interpreted as an implicit recognition of ecotourism's potential to deliver a manifold capitalist fix.

Ecotourism and the Phenomenological Alignment of the Body as a Site of Capitalist Production and Consumption

In this section, we take the preceding analysis one step further by highlighting an additional intriguing dynamic by means of which ecotourism may assist in the quest to overcome limitations to capitalist accumulation, providing what might be called a "bodily fix" (Guthman and DuPuis 2006) to complement the others.[4] In other words, in addition to expanding geographically, ecotourism transforms the human body into an important site of capital accumulation (Harvey 2000; Guthman and DuPuis 2006). This occurs in a number of ways. First, there is the requirement to purchase appropriate equipment to outfit the body for one's excursion. As Brooks (2000: 213) facetiously observes, ecotourists cannot merely interact with "nature" directly but must "master the complex science of knowing how to equip yourself, which basically requires joint degrees in chemistry and physics from MIT." The proper shoes, socks, underwear, pants, shirt, sweater, jacket, hat, scarf, sunglasses, sunscreen, insect repellent, water bottle, headlamp, and backpack—not to mention all of the specialized equipment needed for one's particular pursuit—are required to bring the body into equilibrium with the "natural environment."

In addition to encouraging the consumption of protective and enhancement layers to prepare the body for one's excursion, ecotourism reaches into the body itself as a site of accumulation. In effect, the production of ecotourism as a commodity often requires a high degree of bodily engagement by tourists. Without the active and often strenuous participation of participants, most whitewater paddling trips, for example, would likely not occur at all and would certainly be in much greater risk of entering into dangerous mishap (Fletcher 2010a). Successful whale watching experiences, similarly, require that many tourists overcome the discomforts of motion sickness, learn "how to observe whales at sea," and become skilled at roughing the ocean on a zodiac boat (Neves 2010). In both cases ecotourists actively deploy their bodies to coproduce the very "experience" (i.e., commodity) they consume, in the very location where the experience takes place, although the participatory and capitalist nature of these processes tends to remain invisible to tourists due to fetishization (Neves 2010: 731).

We contend that these processes amount to the phenomenological alignment of the body as a site of capitalist production and consumption (Neves 2009b). By this we mean that in ecotourism, experiences of "being in the world" are scripted and choreographed such that tourists adopt specific kinds of embodiments while they engage in the coproduction of ecotourism commodities, with tour guides frequently acting as coaches, with varying degrees of commandership (Fletcher 2010a). Tourists are told where and when to sit, stand, look, walk, move, stay still, and so on. They are also told which senses to use, how, and when. They are often told what to feel and when. In whitewater rafting, for example, guides commonly manipulate the experience in order to enable passengers to experience the proper level of perceived risk conferring a sense of stimulating excitement without debilitating fear, with the guides providing often explicit instruction concerning how to "correctly" interpret the experience (Fletcher 2010a; Holyfield 1999). In whale watching, there is a privileging of the visual gaze (Urry 2001) over other forms of embodied engagement with whales (instead of, for instance, an auditory perception of whales, attuning to their rhythms and tempos, or paying attention to the experience from the perspective of a more holistic communicative interaction between whale watchers and whale; see Neves 2004, 2006). As noted at the outset, many whale watching tourists are also predisposed to "feel" specific emotions during their encounters with whales due to expectations they have acquired through popular discourses and myths about cetaceans.

The phenomenological alignment of the body as a site of capitalist production and consumption in ecotourism is further extended through the mediation of technology. The production/ consumption of whale watching as commodity is mediated by photo and video cameras, as are whitewater rafting and kayaking trips. For a large percentage of ecotourists a defining activity of whale watching as a commodity is to photograph a whale fluke—some companies will actually refund clients if there are no such photo opportunities on a trip (Neves 2004, 2006). Whitewater rafting clients frequently purchase photographs of themselves in the midst of challenging rapids as documentation of their experience.

In short, we argue that, however subtly, the orchestration of ecotourist experiences amounts to a disciplining of the body (Foucault 1975) whereby ecotourist bodies become sites of capitalist accumulation and tourists become participants in the ongoing cooptation of socio-natures (Hinchliffe 2007) within a neoliberal mode of capitalist conservation (Büscher et al. 2012). In this sense, tourists are simultaneously coproducers and consumers of ecotouristic commodities, while the experience is packaged such that its capitalist nature is fetishized via the construction that the experience actually transcends the shortcomings of capitalism. Ecotourism provides a realm of further accumulation in its commodification of a particular bodily experience achieved during the transitory event of the excursion. In essence, what ecotourism sells most centrally is a particular affective state—excitement, satisfaction, peace, contentment, pleasure, and so forth—attached to the outdoor, generally "wilderness" experience it offers. Commodification of this experience can be seen as yet another attempt to harness crises created by capitalist society as a source of value, promising to compensate for the routinized, alienating nature of most labor within a capitalist mode of production. Ecotourists, indeed, frequently describe their pursuits as an attempt to escape the ostensive monotony, anxiety, dissatisfaction, alienation, sense of fragmentation, and stress of life within modern capitalist society (see, e.g., Arnould and Price 1993; Arnould et al. 1999; Fletcher 2008; Mitchell 1983; Ortner 1999) in pursuit of an experience often characterized as "flow." As Csikszentmihalyi describes:

> Flow refers to the holistic sensation present when we act with total involvement. It is a kind of feeling after which one nostalgically says: "that was fun" or "that was enjoyable." It is the state in which action follows upon action according to an internal logic which seems to need no conscious intervention on our part. We experience it as a unified flowing from one moment to the next in which we are in control of our actions, and in which there is little distinction between self and environment; between stimulus and response; or between past, present, and future. (1974: 58)

Insofar as ecotourists seek to reenchant their lives by engaging in experiences that purport to provide them with deep affective responses, even spiritual cleansing (Arnould and Price 1993; Arnould et al. 1999; Fletcher 2008; Palmer 2004; Stranger 1999), a certain emotional catharsis is part of the package that is offered for sale. Such a "product" is in fact ubiquitous in ecotourism marketing campaigns. Whitewater rafting outfitters commonly promise a transformative if not life-changing experience (Arnould and Price 1993; Arnould et al. 1999; Fletcher 2010a; Holyfield 1999), as evidenced by one prominent outfitter's promotion of its excursion as a "Trip for a Lifetime." Whale watching advertising is so replete with the promise that encounters with whales and dolphins can radically transforms one's life that a new "cathartic healing" industry of swimming with dolphins has been booming for the past decade. Overall, our research shows that tourists seek to reintegrate themselves as "enriched" full persons thus overcoming a sense of fragmentation they feel in their—mostly urban middle class—daily lives within a capitalist society.

Yet this promise of resolution conferring a state of oneness with nature in response to the alienating character of capitalist society is largely an illusory one. The ecotourism experience is a temporary state that invariably returns one to the same everyday life conditions from which one sought to escape. As Mitchell describes, the flow experience is generally fleeting, after which:

> Clarity is replaced with confusion, simplicity with alternatives to be considered, confidence with trepidation, selflessness with self-consciousness. What was moments ago unambiguous now becomes complex; decisions are not clear-cut; the way to go is uncertain. The conditions of the everyday world reimpose themselves on the climber's consciousness. (1983: 168)

In other words, the affective release offered in ecotourism is transitory, and hence rather than delivering an enduring satisfaction of existential angst the experience usually provides merely a "pseudocatharsis" that paradoxically leaves the subject even more dissatisfied through deprivation of the previous stimulation. Yet the fleeting flow experience provides enough pleasure that its subsequent withdrawal inspires a desire for further experience in the hope of recapturing the previous "high" and thereby achieving the enduring resolution thus far denied. In this way, an opportunity for further accumulation is created as tourists seek to reexperience the desired emotional stimulation in search of a continually deferred satisfaction. As the object of this process is an ephemeral affective state that passes quickly with little residual impact on the body, this accumulation process can be virtually infinite, facilitating continual capitalization without significant limit or consequence.

To understand how this dynamic functions, it may help to draw on Slavoj Žižek's idiosyncratic synthesis of Marx and Lacan. Žižek emphasizes the important role of fantasy in sustaining desire for an impossible satisfaction that paradoxically enhances the very desire it inevitably fails to fulfill. As he describes, fantasy "constitutes the frame through which we experience the world as consistent and meaningful" (1989: 138), constructing the ideals that we strive to attain in our own experience and the rewards we believe them to offer. Fantasy thus stimulates desire for what Lacan called *jouissance*, usually translated as "enjoyment" but more properly a mixture of pleasure and pain that promises a satisfaction it can never deliver. This impossible promise ensures, paradoxically, that unresolved desire is sustained rather than resolved, for as Lacan asserted, desire is always at root a desire for desire itself. Hence, "In the fantasy-scene desire is not fulfilled, 'satisfied,' but constituted" (Žižek 1989: 132).

Ecotourism, we assert, tends to operate in just this manner, offering a fantasy of fulfillment that stimulates the desire it promises to resolve while in reality commonly withholding resolution by delivering merely a pseudocatharsis via stimulation of a temporary *jouissance*.

A Whale of a Thing

I (Neves) conducted fieldwork on whale watching in Lajes do Pico from 1998 to 2000 and have been going back frequently since then. I have accompanied endless whale watching trips and interviewed countless tourists during this time. In addition to the ways in which tourists use their bodies to coproduce whale watching commodities, one of the topics that has most interested me in this context is their search for cathartic experiences in whale and dolphin watching. The experience described in the opening vignette represents circa 60 percent of tourists who visit Lajes do Pico (the remainder are local high school students, highly educated tourists, scientific researchers, and health tourists). Despite the fact that many of these tourists take motion sickness medication before embarking on a whale watching trip (as many as 25%) they seek a deeply emotional experience from their encounter with whales. Although the environment of a small zodiac boat is far from intimate, they imagine that their encounter with the whale will provide a "unique one-on-one" experience that will forever change their lives; "that somehow," as one tourist stated, "there will be a connection there that will bring new meaning to life, a clarity that has been missing."

Interestingly, although the majority of these tourists end up seeing only the whale's fluke, and only through the camera's peephole, when I interview them about the experience they do describe it as "magical," "transformative," "cathartic." But they also quickly add that the experience came with limitations. People I interview state that "well … it wasn't all that it could have been … it was great, it was special … but … next time …" and then add a list of conditions that will allow them to improve the experience. These include: "tomorrow I will come back and try the morning trip—I heard whales are more active in the morning"; "next time I am going to try the other boat, I heard you can see much better than from a zodiac"; and "I shouldn't have been on gravol." Although these are all accurate statements on how to improve a whale watching experience, the tourists I have followed on subsequent trips systematically display a pattern of returning with new ideas on how to improve future experiences in order to obtain deeper catharsis ("Next time I will try canoeing among humpback whales in the Antarctic"). There appears to be an escalation of consumptive demand as one pleasurable experience of whale watching increases the threshold of expectations for the next experience. Each experience provides sufficient satisfaction and pleasure to partly fulfill its promises of catharsis, but not enough to be fully meaningful, thus demanding another better fix from a subsequent experience.

Contradictions in Contradictions

In promising a series of fixes for contradictions of capitalist accumulation, ecotourism offers the prospect of continual accumulation without conceivable limits. This prospect, in turn, stimulates further fantasies intrinsic to capitalist ideology, namely the twin promises of accumulation without end and consumption without consequence central to the growing global enthusiasm for employing neoliberal market mechanisms to address the environmental degradation widely seen as exacerbated by capitalism itself (Büscher et al. 2012; Fletcher 2012). In the second fantasy, so-called "ethical consumption" claims to resolve the contradiction between increased consumption and ecological/social crisis by ostensibly linking purchase to social programs that actually redress rather than stimulate such crises (Carrier 2010; Igoe 2010; Igoe et al. 2010; West 2010); while market environmentalism claims to resolve the parallel opposition between economic growth and environmental limits by promoting ostensibly sustainable—even "non-consumptive" (West 2006)—resource exploitation (Brockington et al. 2008; Büscher et al. 2012).

Both of these fantasies are linked to the meta-fantasy at the heart of neoliberalism itself, which Dean, drawing on Žižek, calls the "fantasy of free trade," describing:

> The fantasy of free trade covers over persistent market failure, structural inequalities, the violence of privatization, and the redistribution of wealth to the "have mores." Free trade sustains at the level of fantasy what it seeks to avoid at the level of reality—namely actually free trade among equal players, that is equal participants with equal opportunities to establish the rules of the game, access information, distribution, and financial networks, etc. (2008: 55)

Central to this fantasy, we have shown, is the neoliberal claim that "that capitalist markets are the answer to their own ecological contradictions" (Büscher 2012: 12), and hence that the free market can redress problems of "market failure." This claim is itself contradictory, as researchers increasingly point out (Brockington et al. 2008; Büscher et al. 2012; Fletcher 2012; Fletcher and Breitling 2012; West 2006). Indeed, close investigation demonstrates that ecotourism's claim to resolve contradictions conceals a series of further contradictions inherent in the process of ecotourism development itself.

First, and most significantly, ecotourism's common claim to enhance rather than degrade natural environments belies the significant ecological impacts involved in ecotourism development (Carrier and Macleod 2005; Duffy 2002, 2008; Neves 2010; West and Carrier 2004). Central to this is the fact that ecotourism largely depends on long haul air transport, a significant contributor to the greenhouse gas emissions exacerbating climate change (Carrier and Macleod 2005; Duffy 2008; Hall and Kinnaird 1994). As a result, Hall and Kinnaird (1994: 111) contend, "The extolling of ecotourism development in faraway lands … may thus be viewed as paradoxical"—particularly when ecotourism takes places in destinations, such as small island nations like the Maldives, threatened by climate change itself.

Ecotourism development embodies a number of other evident contradictions as well. As Butcher (2006a, 2006b) points out, while ecotourism claims to constitute a form of development (what West [2006] calls "conservation-as-development"), it really delivers what might be more properly described as "de-development"; that is, the freezing of rural areas in a reified, undeveloped state that precludes the introduction of changes inconsistent with the idealized ecotourism landscape (see also West and Carrier 2004). In addition, while claiming to value traditional knowledge, ecotourism tends to value only that knowledge consistent with its aims; local knowledge conflicting with the interests of ecotourism (advocating resource extraction, for instance) must on the contrary be transformed (Butcher 2006a; Neves 2004).

West and Carrier (2004) identify a further series of contradictions in ecotourism development. First, they highlight ecotourism's "tendency to lead not to the preservation of valued ecosystems but to the creation of landscapes that conform to important Western idealizations of nature through a market-oriented nature politics" (Ibid.: 484). Second, they note an "apparent contradiction between a rhetoric that appreciates and supports exotic local communities and a practice that encourages the socioeconomic values associated with capitalistic individualism" (Ibid.: 485). Third, they highlight a common pressure "towards subordinating concern for environmental conservation and respect for local communities, which ecotourism is said to encourage, to concern for attracting ecotourists and their money" (Ibid.: 491).

Van den Bremer and Büscher (n.d.), finally, add several more contradictions to this discussion. Ecotourism development, they contend, "renders Western culture both the problem and the solution to environmental degradation." This places local stakeholders in a bind in that "when indigenous people 'develop' they become a negative influence to their environment, while this same 'development' in the West has led to the noble ideas of environmentalism and sustainability." Further, the commodification of local landscapes through ecotourism means that "'authentic' and locally particular ecotourism expressions and actor dynamics can simultaneously acquire tendencies that transcend the local and the 'authentic'" by "acquiring global semblances."

These various contradictions are commonly obfuscated via what Carrier and Macleod (2005) call the "ecotourism bubble" and Van den Bremer and Büscher (n.d.) an "ecotourism script." As Carrier and Macleod (2005) point out, ecotourism is commonly marketed as a transparent practice clearly revealing the backstage (MacCannell 1999) infrastructure generally concealed within mass tourism operations (by highlighting, for instance, the social and environmental impacts of tourists' own activities and thereby bursting the "bubble" in which most conventional tourists are immersed during their trips). But in reality ecotourism commonly creates its own tourist bubble by obscuring negative environmental and social consequences in conflict with the virtuous image operators wish to present.

Our own research again illustrates the issue. For instance, most of the tourists who arrived in Lajes do Pico Azores for whale watching between 1989 and 1999 left the tour bus near the village's main pier right by the whalers' museum. The museum was located right beside the most modern whale watching company on the island. The majority of tourists were dazzled by the prospect of the whale watching trip itself, happily shopping while they waited at the company store enjoying its brand-name clothing, postcards, stuffed toys, and so forth. In this bubble of branded whale and dolphin bliss they never noticed the old whale hunters sitting on the benches outside the museum and near the pier who had lost their source of income with the end of whaling in 1983. Ironically, these were the former whalers to whom the government had not provided the opportunity and financial support the aforementioned company had received a few years later to start its own business (Neves 2004, 2006). Most of these ex-whalers had been living on meager early retirement plans since 1983, barely able to make ends meet. From inside their bubble fewer ecotourists would guess the existence of serious tensions in the village between the owner of this whale watching business and local companies trying to develop alternative practices based on what they believed were much sounder and more sustainable relations with cetaceans and a more socially equitable distribution of income (Neves 2004, 2006, 2010).

In addition, and perhaps most important, observations over a period of more than two years indicate that the average marine ecotourist is caught in a bubble in her/his relations with cetaceans, oblivious to potentially damaging environmental effects on at least two levels. First, in many places of the world whale watching clients are expected to act as coenforcers of whale watching rules and regulations. Most of these rules are simple—looking for signs of whale distress such as rapid breathing; refraining from chasing whales; keeping specific distances—but most whale watchers are too preoccupied with securing a good picture, onboard safety, and staving off motion sickness to fulfill this duty. Most simply lack knowledge of cetaceans and are hence unable to enforce cetacean protection rules they have just learned. Second, for the majority of clients the main goal of a whale watching trip is a picture of a fluking whale. Often the best angle to obtain such a picture is right behind the whale. Unless the whale happens to "naturally" decide to dive, however, whale watchers often frighten whales to expedite the process. In the safety of their bubble most tourists are oblivious to the disruption this may cause a whale as they happily click away with their cameras (Neves 2010).

Hence, despite its claim to demystify the tourism experience, ecotourism often functions as a form of commodity fetishism itself (Carrier and Macleod 2005). Van den Bremer and Büscher (n.d.) go further to contend that obfuscation of contradictions within a self-congratulatory "ecotourism script" is in fact essential to ecotourism's success, as admission of negative impacts would compromise operators' ability to present the celebratory image necessary to attract the clients and funders vital for survival. Hence, ecotourism development presents a strong incentive to conceal its inconsistencies beneath a veneer of unequivocal positive benefit to environments and communities alike. In this manner, the neoliberal fantasy of free trade facilitating an infinite process of capital accumulation sustainable along economic, environmental, social, and psychological axes simultaneously is maintained.

Conclusion

We certainly do not intend to deny that ecotourism can at times produce positive results in particular circumstances. Previous research has documented a variety of cases in which ecotourism has in fact contributed to conservation as well as community well-being or empowerment (see, e.g., Almeyda Zambrano et al. 2010; Honey 2008; Krüger 2005; Nyaupane et al. 2006; Stronza 2005, 2010). Neither is this to deny that tourism can at times be employed as an instrument of social justice, even anticapitalist struggle (Higgins-Desbiolles 2006, 2008). Rather, our analysis questions the extent to which ecotourism can truly fulfill its overarching promise to facilitate sustainable development on a global scale by reconciling economic growth with both environmental protection and poverty alleviation within a capitalist framework. As we have shown, although the process of ecotourism development purports to reconcile a number of contradictions intrinsic to capitalist accumulation, the process is itself contradictory in many respects, not least in terms of its ambition to harness the same market mechanisms in large part responsible for many of our social and ecological problems to resolve the very same.

Future research is needed to assess in greater specificity aspects of this analysis presented in largely abstract terms. First, research would be useful to empirically investigate the ways in which ecotourism opens new arenas for capital expansion and accumulation, detailing how this process actually occurs on the ground in particular times and places. Second, study would be valuable to assess how well our analysis of ecotourism as a "bodily fix" resonates with other experience elsewhere. Third, research could explore how contradictions implicit in ecotourism development are "sutured" by the ecotourism script, analyzing the specific discourse employed in this process in particular contexts (Van den Bremer and Büscher n.d.). Finally, analysis might assess the limits to ecotourism's capacity to function as a manifold fix, weighing the benefit derived from this process versus the cost incurred via problems generated in the course of ecotourism development itself. Through this effort, we can gain a more comprehensive understanding of ecotourism's potential to fulfill the lofty promise commonly attributed to it.

■
ROBERT FLETCHER is associate professor of Natural Resources and Sustainable Development in the Department of Environment, Peace and Security at the United Nations-mandated University for Peace in Costa Rica. He has conducted ethnographic research in North, Central, and South America concerning the practice of ecotourism as a strategy for environmental conservation and sustainable development in addition to working for many years

as an ecotourism guide and planner in a variety of locations. He is the author of *Romancing the Wild: Cultural Dimensions of Ecotourism* (forthcoming with Duke University Press).

KATJA NEVES is associate professor of Sociology of the Environment at Concordia University, Montreal, Canada. She currently holds two research grants from the Social Sciences and Humanities Research Council, Canada, and the endorsement of Botanical Gardens Conservation International to investigate the contemporary reinvention of urban Botanical Gardens around the world as agents of biodiversity conservation agents. She is thus also theorizing the emergence of urban socio-natures and the establishment of multistakeholder governmentality within the context of post-2008 austerity discursive economic frameworks.

▪ NOTES

1. Our use of the term "nonhuman natures" follows from contemporary social theory that is sensitive to widespread critique of dichotomous "society/nature" conceptualization. Best articulated in Bruno Latour's book *Politics of Nature* (2004), this critique demonstrates that the ontological distinction between "society" and "nature" is a highly problematic, arbitrary, and hierarchical modern construct. Moreover, the critique reveals that throughout modernity this ontology has constituted the grounds on which the domination and exploitation of "nature" (as well as humans who are deemed to belong to nature, e.g., "tribal peoples") has been politically and ethically justified. Use of terms such as "human natures" and "nonhuman natures," by contrast, evoke the continuum/entanglement, plurality, and heterarchy of "society and nature."
2. http://www.un.org/documents/ecosoc/res/1998/eres1998-40.htm; accessed 12 August 2010.
3. This is certainly not to endorse a functionalist perspective holding that such fixes are the principle reason for tourism's existence, merely that tourism often fulfills such functions as a component of its development. Nevertheless, our research does indicate that the outcomes of ecotourism have been able to offer such fixes, and indeed, that national governments often use the promise of such fixes to justify further investment in and even subsidization of the ecotourism sector (see, e.g., Neves 2004, 2006).
4. This analysis builds on research addressing the role of bodily experience in tourist activity (e.g., Cater and Cloke 2007; Desmond 1999; Graburn 2004; Veijola and Valtonen 2007), responding in part to studies such as Urry's popular *Tourist Gaze* (2001) that neglect to emphasize this important dimension of the experience. Again, this dynamic is not unique to ecotourism yet for purposes of this article we limit our analysis to nature-based pursuits.

▪ REFERENCES

Almeyda Zambrano, Angelica M., Eben N. Broadbent., and William H. Durham. 2010. "Social and Environmental Effects of Ecotourism in the Osa Peninsula of Costa Rica: The Lapa Ríos Case." *Journal of Ecotourism* 9, no. 1: 62–83.

Arnould, Eric J., and Linda L. Price. 1993. "River Magic: Extraordinary Experiences and the Extended Service Encounter." *Journal of Consumer Research* 20: 24–45.

Arnould, Eric J., Linda Price, and Cele Otnes. 1999. "Making Consumption Magic: A Study of White-Water River Rafting." *Journal of Contemporary Ethnography* 28: 33–68.

Bakker, Karen. 2009. "Neoliberal Nature, Ecological Fixes, and the Pitfalls of Comparative Research." *Environment and Planning A* 41: 1781–1787.

Bandy, Joe. 1996. "Managing the Other of Nature: Sustainability, Spectacle, and Global Regimes of Capital in Ecotourism." *Public Culture* 8: 539–566.

Becker, Howard S. 1953. "Becoming a Marihuana User." *American Journal of Sociology* 59: 235–242.

Bellamy Foster, John. 2000. *Marx's Ecology: Materialism and Nature.* New York: Monthly Review Press.

Bianchi, Raoul V. 2005. "Tourism Restructuring and the Politics of Sustainability: A Critical View from the European Periphery (The Canary Islands)." *Journal of Sustainable Tourism* 12, no. 6: 495–529.

Bianchi, Raoul V. 2009. "The 'Critical Turn' in Tourism Studies: A Radical Critique." *Tourism Geographies* 11, no. 4: 484–504.

Braun, Bruce. 2003. "'On the Raggedy Edge of Risk': Articulations of Race and Nature after Biology." Pp. 175–203 in *Race, Nature, and the Politics of Difference,* ed. D.S. Moore, J. Kosek, and A. Pandian. Durham, NC: Duke University Press.

Britton, Stephen G. 1982. "The Political Economy of Tourism in the Third World." *Annals of Tourism Research* 9: 331–358.

Britton, Stephen G. 1991. "Tourism, Capital, and Place: Towards a Critical Geography of Tourism." *Environment and Planning D* 9: 451–478.

Brockington, Dan, and Rosaleen Duffy, eds. 2010. *Antipode* 42, no. 3. Special issue on "Capitalism and Conservation."

Brockington, Dan, Rosaleen Duffy, and Jim Igoe. 2008. *Nature Unbound: Conservation, Capitalism and the Future of Protected Areas.* London: Earthscan.

Brooks, David. 2000. *Bobos in Paradise: The New Upper Class and How They Got There.* New York: Simon & Schuster.

Büscher, Bram. 2010. "Seeking 'Telos' in the 'Transfrontier'? Neoliberalism and the Transcending of Community Conservation in Southern Africa." *Environment and Planning A* 42: 644–660.

Büscher, Bram. 2012. "Payments for Ecosystem Services as Neoliberal Conservation: (Reinterpreting) Evidence from the Maloti-Drakensberg, South Africa." *Conservation and Society* 10, no. 1: 29–41.

Büscher, Bram, Dan Brockington, Jim Igoe, Katja Neves, and Sian Sullivan. 2012. "Towards a Synthesized Critique of Neoliberal Biodiversity Conservation." *Capitalism Nature Socialism* 23, no. 2: 4–30.

Butcher, Jim. 2006a. "Natural Capital and the Advocacy of Ecotourism as Sustainable Development." *Journal of Sustainable Tourism* 14, no. 6: 529–544.

Butcher, Jim. 2006b. "The United Nations International Year of Ecotourism: A Critical Analysis of Development Implications." *Progress in Development Studies* 6, no. 2: 146–156.

Campbell, Colin. 1987. *The Romantic Ethic and the Spirit of Modern Consumerism.* Oxford: Basil Blackwell.

Carrier, James G. 2010. "Protecting the Environment the Natural Way: Ethical Consumption and Commodity Fetishism." *Antipode* 42, no. 3: 672–689.

Carrier, James G., and Donald V.L. Macleod. 2005. "Bursting the Bubble: The Socio-Cultural Context of Ecotourism." *Journal of the Royal Anthropological Institute* 11: 315–334.

Castree, Noel. 2008. "Neoliberalising Nature: The Logics of Deregulation and Reregulation." *Environment and Planning A* 40: 131–152.

Castree, Noel. 2010. "Crisis, Continuity and Change: Neoliberalism, the Left and the Future of Capitalism." *Antipode* 41(S1): 185–213.

Cater, Carl, and Paul Cloke. 2007. "Bodies in Action: The Performativity of Adventure Tourism." *Anthropology Today* 23, no. 6: 13–16.

Cater, Erlet. 2006. "Ecotourism as a Western Construct." *Journal of Ecotourism* 5, nos. 1–2: 23–39.

Covington, Dennis. 1995. *Salvation on Sand Mountain: Snake Handling and Redemption in Southern Appalachia.* New York: Penguin.

Crapper, Mary M. 1998. "From Hunters to Guides: How Some Native Communities Have Profited from Ecotourism." *Contact Peru* 3: 20–21.

Csikszentmihalyi, Mihaly. 1974. *Flow: Studies in Enjoyment.* Public Health Service Grant Report No. RO1HM 22883-02.

Davis, Susan G. 1997. *Spectacular Nature: Corporate Culture and the Sea World Experience.* Berkeley: University of California Press.

Dean, Jodi. 2008. "Enjoying Neoliberalism." *Cultural Politics* 4, no. 1: 47–72.

Desmond, Jane C. 1999. *Staging Tourism: Bodies on Display from Waikiki to Sea World.* Chicago: University of Chicago Press.

Doane, Molly. 2010. "Maya Coffee: Fair Trade Markets and the 'Social Fix.'" Paper presented at the American Anthropological Association Annual Conference, New Orleans, LA, 17–21 November.

Dressler, Wolfram, and Robin Roth. 2010. "The Good, the Bad, and the Contradictory: Neoliberal Conservation Governance in Rural Southeast Asia." *World Development* 39, no. 5: 851–862.

Duffy, Rosaleen. 2002. *Trip Too Far: Ecotourism, Politics, and Exploitation.* London: Earthscan.

Duffy, Rosaleen. 2008. "Neoliberalising Nature: Global Networks and Ecotourism Development in Madagascar." *Journal of Sustainable Tourism* 16, no. 3: 327–344.

Duffy, Rosaleen. 2010. *Nature Crime.* New Haven, CT: Yale University Press.

Duffy, Rosaleen. 2012. "The International Political Economy of Tourism and the Neoliberalisation of Nature: Challenges Posed by Selling Close Interactions with Animals." *Review of International Political Economy,* doi: 10.1080/09692290.2012.654443.

Duffy, Rosaleen, and Loraine Moore. 2010. "Neoliberalising Nature? Elephant-back Tourism in Thailand and Botswana." *Antipode* 42, no. 3: 742–766.

Escobar, Arturo. 1995. *Encountering Development: The Making and Unmaking of the Third World.* Princeton: Princeton University Press.

Fletcher, Robert. 2008. "Living on the Edge: The Appeal of Risk Sports for the Professional Middle Class." *Sociology of Sport Journal* 25, no. 3: 310–330.

Fletcher, Robert. 2009. "Ecotourism Discourse: Challenging the Stakeholders Theory." *Journal of Ecotourism* 8, no. 3: 269–285.

Fletcher, Robert. 2010a. "The Emperor's New Adventure: Public Secrecy and the Paradox of Adventure Tourism." *Journal of Contemporary Ethnography* 39, no. 1: 6–33.

Fletcher, Robert. 2010b. "Neoliberal Environmentality: Towards a Poststructuralist Political Ecology of the Conservation Debate." *Conservation and Society* 8, no. 3: 171–181.

Fletcher, Robert. 2011. "Sustaining Tourism, Sustaining Capitalism? The Tourism Industry's Role in Global Capitalist Expansion." *Tourism Geographies* 13, no. 3: 443–461.

Fletcher, Robert. 2012. "Using the Master's Tools? Neoliberal Conservation and the Evasion of Inequality." *Development and Change* 43, no. 1: 295–317.

Fletcher, Robert, and Jan Breitling. 2012. "Market Mechanism or Subsidy in Disguise? Governing Payment for Environmental Services in Costa Rica." *Geoforum* 43, no. 3: 402–411.

Foucault, Michel. 1975. *Discipline and Punish: The Birth of the Prison.* New York: Random House.

Gibson, Chris 2009. "Geographies of Tourism: Critical Research on Capitalism and Local Livelihoods." *Progress in Human Geography* 33, no. 4: 527–534.

Graburn, Nelson. 2004. "Secular Ritual: A General Theory of Tourism." Pp. 23–34 in *Tourists and Tourism,* ed. S.B. Gmelch. Long Grove, IL: Waveland Press.

Guthman, Julie, and Melanie DuPuis. 2006. "Embodying Neoliberalism: Economy, Culture and the Politics of Fat." *Environment and Planning D: Society and Space* 24: 427–448.

Hall, Derek R., and Vivian Kinnaird. 1994. "Ecotourism in Eastern Europe." Pp. 111–136 in *Ecotourism: A Sustainable Option?* ed. E. Cater and G. Lowman. Chichester, UK: Wiley.

Harvey, David. 1989. *The Condition of Postmodernity: An Inquiry into the Origins of Cultural Change.* Oxford: Basil Blackwell.

Harvey, David. 2000. *Spaces of Hope.* Berkeley: University of California Press.

Harvey, David. 2006. *Spaces of Global Capitalism: A Theory of Uneven Geographical Development.* London: Verso.

Heynen, Nik, James McCarthy, Paul Robbins, and Scott Prudham, eds. 2007. *Neoliberal Environments: False Promises and Unnatural Consequences.* New York: Routledge.

Higgins-Desbiolles, Freya. 2006. "More Than an 'Industry': The Forgotten Power of Tourism as a Social Force." *Tourism Management* 27: 1192–1208.

Higgins-Desbiolles, Freya. 2008. "Justice Tourism and Alternative Globalization." *Journal of Sustainable Tourism* 16, no. 3: 345–364.

Hinchliffe, Steve. 2007. *Geographies of Nature: Societies, Environments, and Ecologies.* Waterloo: Wilfrid Laurier Press.

Holyfield, Lori. 1999. "Manufacturing Adventure: The Buying and Selling of Emotions." *Journal of Contemporary Ethnography* 28, no. 1: 3–32.

Honey, Martha. 2008. Eco*tourism and Sustainable Development: Who Owns Paradise?* 2nd ed. Washington, DC: Island Press.

Igoe, Jim. 2010. "The Spectacle of Nature in the Global Economy of Appearances: Anthropological Engagements with the Spectacular Mediations of Transnational Conservation." *Critique of Anthropology* 30, no. 4: 375–397.

Igoe, Jim, and Dan Brockington. 2007. "Neoliberal Conservation: A Brief Introduction." *Conservation and Society* 5, no. 4: 432–449.

Igoe, Jim , Katja Neves, and Dan Brockington. 2010. "A Spectacular Eco-Tour around the Historic Bloc: Theorising the Convergence of Biodiversity Conservation and Capitalist Expansion." *Antipode* 42, no. 3: 486–512.

Klein, Naomi. 2007. *The Shock Doctrine: The Rise of Disaster Capitalism.* New York: Metropolitan Books.

Krüger, Oliver. 2005. "The Role of Ecotourism in Conservation: Panacea or Pandora's Box?" *Biodiversity and Conservation* 14: 579–600.

Langholz, Jeffrey A., and James P. Lassoie. 2001. "Perils and Promise of Privately Protected Areas." *BioScience* 51, no. 12: 1079–1085.

Lash, Scott, and John Urry. 1987. *The End of Organized Capitalism.* Cambridge: Polity Press.

Latour, Bruno. 2004. *Politics of Nature: How to Bring the Sciences into Democracy.* Cambridge, MA: Harvard University Press.

Leahy, Stephen. 2008. "Extinction Tourism: See It Now Before It's Gone." http://stephenleahy.net/2008/01/18/extinction-tourism-see-it-now-before-its-gone/ (accessed 8 January 2012).

MacCannell, Dean. 1999. *The Tourist: A New Theory of the Leisure Class.* 2nd ed. Berkeley: University of California Press.

Marx, Karl. 1973. *Grundrisse: Foundations of the Critique of Political Economy.* Harmondsworth: Penguin.

McCarthy, James, and Scott Prudham. 2004. "Neoliberal Nature and the Nature of Neoliberalism." *Geoforum* 35, no. 3: 275–283.

Mitchell, Richard G., Jr. 1983. *Mountain Experience: The Psychology and Sociology of Adventure.* Chicago: University of Chicago Press.

Mowforth, Martin, and Ian Munt. 2008. *Tourism and Sustainability: New Tourism in the Third World.* 3rd ed. London: Routledge.

Munt, Ian. 1994. "Eco-Tourism or Ego-Tourism?" *Race and Class* 36: 49–60.

Neves, Katja. 2004. "Revisiting the Tragedy of the Commons: Whale Watching in the Azores and its Ecological Dilemmas." *Human Organization* 63, no. 3: 289–300.

Neves, Katja. 2006. "Politics of Environmentalism and Ecological Knowledge at the Intersection of Local and Global Processes." *Journal of Ecological Anthropology* 10: 19–32.

Neves, Katja. 2009a. "The Sacredness of Human-Cetacean Unities: Towards a Non-Mechanical Non-Transcendental Ethnographic Approach." Paper presented at the American Academy of Religion Conference, Montreal, Canada, 7–10 November.

Neves, Katja. 2009b. "Exaggerated News of a Timely Demise: IUCN's Conservation Categories and the Possibility for Neoliberal Rebirth." Paper presented at the American Anthropological Association Annual Conference, Philadelphia, PA, 2–6 December.

Neves, Katja. 2010. "Cashing in on Cetourism: A Critical Engagement with Dominant E-NGO Discourses on Whaling, Cetacean Conservation, and Whale Watching." *Antipode* 42, no. 3: 719–741.

Neves, Katja. Under Review. "The Politics of Multi-Faceted World Heritage: Pico Vineyard's Cultural Landscape." *American Anthropologist.*

Neves, Katja. n.d. '*A Whale of a Thing': The Political Ecology of Whaling and Whale Watching and in a World of Neoliberal Conservation.* Unpublished manuscript.

Neves, Katja, and Jim Igoe. 2012. "Accumulation by Dispossession and Uneven Development: Comparing Recent Trends in the Azores and Tanzania." *Journal of Economic and Social Geography* 103, no. 2: 164–179.

Nyaupane, Gyan P., Duarte B. Morais, and Lorraine Dowler. 2006. "The Role of Community Involvement and Number/Type of Visitors on Tourism Impacts: A Controlled Comparison of Annapurna,
Nepal, and Northwest Yunnan, China." *Tourism Management* 27: 1373–1385.

O'Connor, James. 1988. "Capitalism, Nature, Socialism: A Theoretical Introduction." *Capitalism Nature
Socialism* 1, no. 1: 11–38.

O'Connor, James. 1994. "Is Sustainable Capitalism Possible?" Pp. 125–137 in *Food for the Future: Conditions and Contradictions of Sustainability,* ed. P. Allen. New York: Wiley-Interscience.

O'Connor, Martin. 1994. "On the Misadventures of Capitalist Nature." Pp. 125–151 in *Is Capitalism Sustainable?* ed. M. O'Connor. New York: Guilford.

Ortner, Sherry B. 1999. *Life and Death on Mt. Everest: Sherpas and Himalayan Mountaineering.* Princeton: Princeton University Press.

Palmer, Catherine. 2004. "Death, Danger, and the Selling of Risk in Adventure Sport." Pp. 55–69 in
Understanding Lifestyle Sports, ed. B. Wheaton. New York: Routledge.

Poon, Auliana. 1993. *Tourism, Technology, and Competitive Strategies.* Wallingford: CABI.

Robbins, Paul, and Alistair Fraser. 2003. "A Forest of Contradictions: Producing the Landscapes of the
Scottish Highlands." *Antipode* 35, no. 1: 95–118.

Smith, Neil. 2007. "Nature as Accumulation Strategy." *Socialist Register* (January): 1–36.

Stranger, Mark. 1999. "The Aesthetics of Risk: A Study of Surfing." *International Journal for the Sociology
of Sport* 34, no. 3: 265–276.

Stronza, Amanda. 2005. "Hosts and Hosts: The Anthropology of Community-Based Ecotourism in the
Peruvian Amazon." *Napa Bulletin* 23: 170–190.

Stronza, Amanda. 2007. "The Economic Promise of Ecotourism." *Journal of Ecotourism* 6, no. 3: 210–230.

Stronza, Amanda. 2010. "Commons Management and Ecotourism: Ethnographic Evidence from the
Amazon." *International Journal of the Commons* 4, no. 1: 56–77.

Sullivan, Sian. 2006. "The Elephant in the Room? Problematising 'New' (Neoliberal) Biodiversity Conservation." *Forum for Development Studies* 33, no. 1: 105–135.

Sullivan, Sian. 2009. "Green Capitalism, and the Cultural Poverty of Constructing Nature as Service Pro-
vider." Pp. 255–272 in *Upsetting the Offset,* ed. S. Böehm and S. Dabhi. London: MayFly Books.

Tambiah, Stanley J. 1990. *Magic, Science, Religion, and the Scope of Rationality.* Cambridge: Cambridge
University Press.

The International Ecotourism Society. 2006. "The Global Ecotourism Fact Sheet." Washington, DC.

United Nations World Tourism Organization (UNWTO). 1998. *Ecotourism, Now One-Fifth of Market.*
Madrid.

Urry, John. 2001. *The Tourist Gaze.* 2nd ed. Thousand Oaks, CA: Sage.

Van den Bremer, Rene, and Bram Büscher. n.d. "The Politics of Sustainable Community Tourism in
Ghana and the 'Ecotourism Script.'" Unpublished manuscript.

Veijola, Soile, and Anu Valtonen. 2007. "The Body in Tourism Industry." Pp. 13–31 in *Tourism & Gender: Embodiment, Sensuality and Experience,* ed. A. Pritchard, N. Morgan, and I. Ateljevic. London:
CABI.

Vivanco, Luis A. 2001. "Spectacular Quetzals, Ecotourism, and Environmental Futures in Monte Verde,
Costa Rica." *Ethnology* 40, no. 2: 79–92.

Vivanco, Luis A. 2006. *Green Encounters: Shaping and Contesting Environmentalism in Rural Costa Rica.*
New York: Berghahn Books.

Weber, Max. 1930. *The Protestant Ethic and the Spirit of Capitalism.* New York: Routledge.

West, Paige. 2006. *Conservation is Our Government Now: The Politics of Ecology in Papua New Guinea.*
Durham, NC: Duke University Press.

West, Paige. 2010. "Making the Market: Specialty Coffee, Generational Pitches, and Papua New Guinea."
Antipode 42, no. 3: 690–718.

West, Paige, and James C. Carrier. 2004. "Ecotourism and Authenticity: Getting Away From it All?" *Current Anthropology* 45, no. 4: 483–498.

Žižek, Slavoj. 1989. *The Sublime Object of Ideology.* London: Verso.

From a Blind Spot to a Nexus

Building on Existing Trends in Knowledge Production to Study the Copresence of Ecotourism and Extraction

Veronica Davidov

ABSTRACT: Ecotourism is primarily perceived and studied as an alternative to resource extraction, even though increasingly the two coexist side by side in a nexus. This article investigates how such instances of copresence are marginalized in literatures about ecotourism and extraction, constituting a "blind spot" in academic literature. An extensive literature review focuses on the existing knowledge trends and paradigms in the production of knowledge about ecotourism and extraction, and analyzes whether they contribute to the "blind spot" or can be mobilized by the nexus perspective. Finally, the article briefly outlines two methodological approaches for studying ecotourism and extraction as a nexus.

KEYWORDS: ecotourism, methodology, nexus, resource extraction, production of knowledge,

Ecotourism in the same place as oil pipes or open pit mines sounds like a strange, contradictory proposition. And, thanks to ongoing new technologies and geographies of the neoliberalization of nature, this "contradiction" abounds in empirical reality: the "protected" Campo Ma'an National Park is being developed for ecotourism by the Cameroonian Ministry of Tourism, 6 miles from the Chad-Cameroon oil pipeline; there are indigenous ecolodges right in the middle of Ecuador's infamous "oil patch"; in Northern Russia there are lakefront ecodestinations literally minutes away from dimension stone quarries.[1]

How can we understand and study these seemingly contradictory situations where ecotourism and natural resource extraction occur side-by-side, sometimes even supported by the same institutions? So far, ecotourism is primarily perceived and studied as an alternative to resource extraction, while studies of resource extraction generally do not include ecotourism projects that may exist in the vicinity of extraction sites. Existing academic and policy literatures privilege oppositions and transitions between "sustainable" and "unsustainable" development over congruencies and synergies, which could reveal the uncertainties, contradictions, and fluidities inherent in this polarization. Because of this existing framing bias, the phenomenon of ecotourism in areas concurrently affected by extraction industries (oil production, mining, logging), remains understudied, even though for a range of reasons such a scenario is increasingly common in resource-rich counties of the Global South.

In general, issues around constructions and contestations of "nature" and "natural resources" have only gained traction in the social sciences in the recent years, from a "reimagined" politi-

Environment and Society: Advances in Research 3 (2012): 78–102 © Berghahn Books
doi:10.3167/ares.2012.030106

cal ecology (Biersack 2006) to the "materiality turn," to articulations of new resource geographies (Bakker and Bridge 2006). Actors across policy and academic arenas engage in debates about the ideologies, policies, and practices of the projects of "development" and nature's role in it, as well as the various forms of environmental governance. Both ecotourism and resource extraction have been central to the debates and discussions—both are about what is defined as a resource, how it is governable (or ungovernable), and how it is mobilized in the service of "development." Conventionally, resource extraction and ecotourism map onto "opposite" discourses of development and resource governance within the symbolic universe that operates through reified dichotomies like "exploitation versus conservation," "commodification versus preservation" and (sometimes) "corporations versus the state" or "corporations versus environmental foundations." Of course, these binaries do not correspond to the complexities of real practices around commodification of nature, but they are successfully reinforced in production of knowledge practices, particularly in institutional and policymaking settings. Thus, as Igoe and Brockington wrote of spaces of knowledge production and expertise exchange pertaining to international bioconservation: "the term 'neoliberal' is not commonly invoked at conferences, on email lists, or in professional journals of biodiversity conservation. To the extent that it is discussed, the suggestion is that international conservation represents a bulwark against neoliberalism by protecting our planet's ecosystems from the advance of free-market capitalism" (2007: 433). At the same time, extraction enterprises try to "green" themselves through publicity-generating environmental impact assessments, projects like "green mining initiatives," and other initiatives designed to display corporate virtue and general assimilation to the narratives and semiotics of "compassionate," "humane," or "green" capitalism.

There is a body of critical literature that problematizes these dichotomies and knowledge practices. Much of this literature is about macroprocesses—a scale necessary for a systematic examination of fundamental economic, ecological, and cultural paradoxes of global and multisited phenomena like "neoliberal conservation" or "sustainable development." The broad field of practices of neoliberalization of nature and "neoliberal conservation" has been defined and examined by the likes of Igoe (2003), Sullivan (2006), Heynen et al. (2007), Brockington et al. (2008), Castree (2008), Brockington and Duffy (2010), Büscher (2010), while scholars like Tsing (2004), Vivanco (2006), and West (2006a) shifted between scales showing how processes of globalization and neoliberalization make local natures.

But even as the lines between extraction and conservation are revealed to be more complex and nebulous than the aforementioned dichotomies (still powerful in shaping public opinion and policy rhetoric) would suggest, the notion of resource extraction and ecotourism, as iconic practices within these domains, coexisting literally side-by-side, is underimagined, underrecognized, and understudied. Why do such dichotomies become harder to challenge when it comes to particular material practices in concrete spaces? When I was first intrigued by the copresence of ecotourism and extraction, which I observed over the course of my fieldwork, and wrote a grant proposal designed to systematically study this phenomenon, one of the reviewers—clearly familiar with the broad theoretical literature on "neoliberal conservation"—nevertheless wrote: "I am rather dubious about this [project]. I find it hard to imagine that landscapes ravaged by extraction of oil or other sub-surface minerals are commonly being selected as sites for ecotourism development."[2] This response helped convince me that even in the academic community certain assumptions about what extraction or ecotourism "do" to landscapes contribute to the seeming impossibility of contemplating them as cooccurring practices. These assumptions concern what constitutes an "extractive industry" and the "visibility" (or invisibility) of particular impacts on nature (while the oil sands infrastructure in Northern Canada is at odds with the aesthetics of a landscape constructed by ecotourism fantasies, in the villages connected by the

Chad-Cameroon pipeline, the only outward sign of the oil flow is a small sign indicating the location of the subterranean pipeline, and the effects of the oil industry are not aesthetic in the same way). But another part of it, I contend, has to do with the conventions and practices of knowledge production about these particular industries, and that is what inspired this review. Over the course of this article, I look, in turn, at the key trends in the literatures on "ecotourism" and "extraction" and see how they construct their subjects, and what kinds of analysis they practice and model for other researchers. Do these epistemological practices detect and analyze extraction and ecotourism when they appear in the same spot on the map? Or do they contribute to decoupling them epistemologically as they are already decoupled ideologically?

I introduce the concept of a nexus in my title: I imagine a nexus as a space of engagement, an arena of entanglement, both in terms of production of knowledge, and in terms of competing or complementary material practices being enacted within a specific landscape. The foundational body of literature I reference already engages the macrocontours of this nexus, its ontological dimensions, by showing how the purported dichotomies between capitalism and conservation break down, and how the two converge. Certain works go as far as identify the institutional dynamics and forces that enable and constitute the nexus—for instance, McDonald's (2010) work on conservation industries finding themselves in bed with the "devil" of "private sector" industries with direct ties to extractive initiatives, or Corson's (2010) inquiry into interorganizational relations that facilitate the neoliberal conservation regime.

For the purposes of this article, I am interested in the copresence of ecotourism and extraction as material practices that coexist geographically in such a way that they simultaneously engage and impact the same community (or communities), and that they would both be salient factors in an ecosystem-scale analysis (depending on the "ecosystem" in question, the distances may seem greater than in a "community-scale analysis" but the effect no less significant). Although "community" and "ecosystem" are nebulous and fluid concepts for a number of reasons, they are both concepts that are frequently and successfully mobilized as units of analysis or sites of impact in debates about valuation and use(s) of nature. They both index a form of connection that is immediate, visceral, and visible, or at least clearly evident because of the imaginaries of "place" with which they are associated. They are the kinds of spaces within which it becomes challenging to imagine the coexistence of two industries that are so iconically divergent and opposite, that stand for and symbolize such different approaches to both "development" and human-nature relations. One could write a separate article on imaginaries of places and scales, that would explore why practices that are clearly linked on the scales of institutions or ideologies are difficult to imagine as co-existing in the same physical space—but that is not my task here. I note this as context for my working definition of the nexus as possessing the contours of close geographical proximity, although, of course, this nexus exists across multiple scales. That is to say that the scale I am interested in is absolutely not the only one through which the nexus is evident; but it is one, I contend, where it is most underrecognized. Although a full, detailed typology of the kinds of ways in which extraction and ecotourism are linked in such places is beyond the scope of this article, the copresence can occur as a result of top-down facilitated converge, or oppositional strategies or tactics. The first type of linkage is often connected to "green capitalism" or corporate social responsibility (CSR) ventures, where "extractive development" and "sustainable development" are imagined to be integrated. The second type, frequently facilitated by local actors who ally with organizations that help them start ecotourism projects, often functions as a kind of praxis critique of extraction where what Agraval (2005) would call "environmentality" is mobilized against what could be termed "resource developmentality." But such oppositional dynamics are complex, with strategic adoptions of particular forms of resource management to send a message that the two approaches are incompatible resulting in their prolonged coexistence (and at least technical compatibility), with what are often the same sets of actors brokering local interests in ecotourism with the extraction industries. In other instances, local actors may not see the activities as paradoxical, and desire to be equally involved in both (Smith 2012).

A place-based nexus can also have temporal aspects—as happens in the cases where extraction activities and ecotourism activities are simultaneously envisioned as successive "development" phrases of a particular location—increasingly, as remediation plans are mandated for successful concession bids, ecotourism and extraction can become conceptualized as sequential, planned regimes of commodifying nature. This sort of management regime both challenges and subverts the dominant imaginary of the sequential-temporal relationship between ecotourism and extraction, where the shift is conceptualized as a result of a tension between alternative development paradigms, rather than a part of a premeditated plan.

Because this is a literature review, rather than an ethnographic, or even a comparative article, I cannot focus in-depth on the specific locations where this cooccurrence takes place, although I note a number of them at the very beginning (see also note 1). These locations, I argue, are places where the logic of neoliberal conservation manifests as material practices. By reviewing the bodies of literature that shape the standards and the vanguards of the epistemologies that bear on those practices, I can hopefully contribute to a larger body of literature, the one dedicated to dereifying and challenging the false dichotomies I invoke at the beginning of this introduction. What follows are reviews of knowledge production trends in the study of ecotourism and resource extraction. As I review these texts, I analyze the trends in questions, frameworks, and research designs, and assess whether the existing conventions can be mobilized for the study of the convergence of ecotourism and extraction as grounded practices and as macroprocesses, or whether they downright impede that kind of examination, producing a blind spot.

The Production of Knowledge about Ecotourism

In lieu of launching into a full review of trends and topics in ecotourism literature, I want briefly to reflect on Weaver and Lawton's (2007) "state of the research in ecotourism" piece. Weaver and Lawton, in their synthesis of 20 years of ecotoursim research, identify and discuss a wide range of categories of research undertaken (or understudied). Although their goals differ from mine, from a practical standpoint their review is highly instrumental, and rather than recapitulating it, I instead use it as a starting basis to comment on the trends they identify, then fill in the "blanks" of their overview, while discussing how these broad trends relate to the absence and the possibility of considering ecotourism and extraction together, as a nexus site of production of knowledge.

Supply and Demand

Weaver and Lawton start out with what they call "supply"—supply of ecotourism itself, that is. They engage with conventions and tensions in the definitions of ecotourism, starting with Fennell's (2001) 85 definitions of ecotourism. They review the literature that has, over the years, attempted to establish a consensus around the definition and the parameters of the practice (Blamey 1997, 2001; Weaver 2005), and in the process still has to reckon with debates of inclusion and exclusion of categories—do "controversial" activities like "trophy hunting" (Novelli et al. 2006) or "natural habitat" zoos (Ryan and Saward 2004) fit the definition? The "what" is immediately followed by the "where" as Weaver and Lawton survey the literature on the "venues"—the spaces and locations of ecotourism. They note that in academic literature on the subject case studies from the so-called "least developed countries" dominate—"perhaps in recognition of the degree to which ecotourism can potentially serve as a vehicle for economic development in the area" (2007: 1170). They comment that relatively little research is undertaken on the ecotourism "industry." Perhaps, they speculate, "because LDC sites that dominate the literature tend to follow the community-based model of service provision that is largely external in this industry" (Ibid.: 1171). On this I want to challenge their point and note that the "community-based model" they reference is in many ways an "industry standard," at least discursively, and any separation between it as an aspirational "model" and a supposedly distinct, firmly demarcated "ecotourism industry" collapses in the face of first "practitioner" and "practitioner-oriented" academic literature that stresses the "community-based model" (Honey 2002; Okazaki 2008). Additionally, ethnographic evidence and critiques of NGO-facilitated, show that "community-based models" struggle, among other things, with gaps between outsider and local definitions of "community" (Chernela 2005a) and misrepresentations of community politics (Belsky 1999). Unsurprisingly, venue-focused literature tends to be case study-based, as exemplified by the examples Weaver and Lawton provide. This focus within this category is largely on "successes" and "failures" of ecotourism, assessed in financial and educational terms. This is also unsurprising, given that ecotourism is a institutional homunculus, launched as a global aspirational enterprise promising to integrate conservation and development, through a sequence of dedicated institutional events (like the Rio Earth Summit and the International Year of Ecotourism), created for efficacy and results, and designed to be assessed, audited, and improved upon.

Unlike resource extraction, which is a multifaceted phenomenon that different types of institutions attempt to regulate and harness, ecotourism, it can be argued, is an institutionally created phenomenon, that has never existed outside of the institutions and bureaucracies of "development," with resultant fairly standardized rhetoric and framework of indicators and "good practices" and "successes" against which any incarnation of an ecotourism venture can theoretically be measured by the people whose literal job it is to do so. Arguably, it is this ontogenesis of "ecotourism" as an institutional phenomenon that is responsible for the dominance of the "successes"- and "failures"-oriented assessments in ecotourism literature. I contend this a priori institutional nature of ecotourism, which promotes a culture of assessment/knowledge production around its "successes" and "failures" is part of how the blind spot I am investigating is produced. If an ecotourism project is built with oil money in close proximity to an oil extraction/transport operation, what does that mean for how the success of it should be measured? If the metrics for success are decontextualized from the extraction going on in close proximity to ecotourism, is "venue-scale" analysis a deterritorialized one, in a sense that it is removed from the actual ecological and geographical ongoings on the territory in question, and is thus guilty of "virtualism," which Carrier and Miller (1998) define as an attempt to make the real world conform to an abstract model of itself?

After supply, Weaver and Lawton tackle demand—this section of their article tries to identify and study ecotourism as a distinct set of consumers, meta-analyzing both the critical literature engaging with the methodologies and scope of such research (Wight 2001) and the various "reports" on the profiles of the "typical" ecotourist—one characterized by "higher levels of education and income," environmental "awareness" compared to "mainstream" consumers, and "disproportionately originat[ing] in MDCs" (Eagles and Cascagnette 1995; Wight 1996). Although actor-level analysis, which is so common for this literature sector is necessary to the study of any system, it can be argued that the dominance of this framework and the research direction it ensures, detracts from a broader reconceptualization of the ecotourism "market" in a more general sense, beyond the "greenies" with backpacks who can be located (and studied) at specific ecotourism venues. Perhaps another set of "consumers" who should be identified and theorized as such are investors who want to engage with a "green" economy and "green" forms of development, or the "parallel" sector of consumers of fair trade products that are often manufactured within the framework of ecotourism endeavors. Ecotourism "consumers" are not necessarily people in search of or physically going to ecotourism destinations—they can also be individual or corporate actors in the market for a reassuring rhetoric of green legitimacy that different forms of association with ecotourism projects (not necessarily direct participation) can provide. In fact, such a form of "demand" is one of the driving mechanisms for the seemingly paradoxical simultaneous conservation/extraction initiatives under discussion, as in one scenario, such a nexus arises as a "corporate social responsibility" practice. What is interesting about such practices is that the easily recognizable way they operate often involves great distances—for example, sovereign wealth funds enriched by domestic extraction in the Global North being used to fund conservation initiatives, including ecotourism, in the Global South. But ultimately, an oil consortium sponsoring a national park with an ecotourism project blueprint minutes away from an oil pipeline follows the same logic, only putting these activities into the same temporal and spatial context, where they become more difficult to reconcile.

An inquiry into the institutional context of ecotourism could provide great insights into how ecotourism and extraction come to exist in the same rainforests and along the same rivers—after all, the phenomenon that may appear initially puzzling or contradictory "on the ground" may become very clear when the financing and the institutional convergences between the forces that facilitated the emergence of the two endeavors side-by-side are illuminated. But, historically, few studies have addressed this topic—as Weaver and Lawton note, "despite the essential nature of this research to the management of the ecotourism experience, almost none of the empirical studies has been undertaken by tourism specialists or is found in specialized tourism journals" while the studies that do exist appear almost exclusively in *Biological Conservation* (2007: 1173). Fortunately, this is a growing field of inquiry, including texts like West and Carrier (2004), Duffy (2008), Fletcher (2009), Brondo and Bown (2011), emerging as part of the broader production of knowledge on financialization and neoliberalization of nature.

"Impactology": Ecotourism

Finally, Weaver and Lawton make note of the broad category of "impacts" (ecological, economic, sociocultural). This is an extensive category, as this "impactology" approach has always been important in cross-disciplinary ecotourism studies. Ecotourism has often been viewed and studied as a "change" mechanism, or an introduction of an external force into study sites. Consequently, every discipline could assess some aspect of the ensuing dynamics—impacts on the ecosystem or the "wildlife" (see Buckley's 2009 review of this literature), including waste (Meletis and Campbell 2009), the local prosperity (Stronza 2010; Stronza and Gordillo 2008; Wunder 2000), social forms, political organization, religious practices, etcetera (Bruner 2005; Cheong and Miller 2000; Davidov 2008; Hutchins 2007). One aspect of "impactology" mentioned perhaps too briefly in Weaver and Lawton's review is the impact of ecotourism on ecotourists over the long term. Stronza, in her review of tourism literature, wrote: "Less effort has been invested in analyzing the incentives ecotourism offers to tourists to change their own perspectives and behaviors…despite the fact that a significant goal of ecotourism is to raise environmental and cultural awareness among tourists" (2001: 77). A decade later, this has been a growing area of inquiry (Hughes et al. 2011; Powell and Ham 2008), although still understudied compared to research on "the tourees."

In general, an "impactology" framework could be well-mobilized for an empirical study of the coexistence of extraction and ecotourism, but the articulation of impact rubrics would have to be dynamic and committed to ecosystem-scale/place-oriented analyses, rather than project-oriented analysis or analysis focused on one particular sphere of investigated impact, such as economic prosperity or non-local categories like "community empowerment" or "capacity-building" (Laverack and Thangphet 2009; Scheyvens 1999). Any productive refashioning of the "impactology" approach would have to be more integrative and context-grounded, and less invested in segmentation between "real" ecotourism and more "questionable" activities, as that distinction in and of itself produces a blind spot for the political economy and political ecology of even the most "textbook" ecotourism ventures. Weaver and Lawton point out that there is an "external environments" trend of analysis that focuses on the "blurring" between ecotourism and "mainstream" tourism. They write: "Less understood is the relationship between ecotourism and extractive activities such as recreational hunting and fishing, assuming that the latter do not qualify as ecotourism" (2007: 1175). It is interesting (and telling) that the extractive activities like mining, oil extraction, and logging are not even on this research map, while their very convergence and complex relationship with ecotourism could be the basis of a very different "impactology" approach. One might argue that hunting and fishing are more ambiguous categories that could potentially be incorporated into ecotourism, unlike mineral extraction, but one only needs to look at the conversion of diamond mining and ecotourism in places like Yellowknife, Canada and the De Beers' Namaqualand mine in South Africa—a conversion pivoting around the symbolic and material figure of the diamond, which is made to represent simultaneously pure nature and the wealth that can be achieved through mining—to see that with the proper "amalgam" objects or activities the boundaries between ecotourism and extraction are also nebulous and fluid.

Critiques

What about ecotourism literature in the years since Weaver and Lawton's overview, or the literature sectors not included in their article? Although the general trends noted by Weaver and Lawton might be phrased somewhat differently and grouped in somewhat different categories, research priorities articulated by ecotourism scholars in the recent years echo the ones identified by the two scholars. In a more recent overview, Buckley (2009: 643) identifies "future research priorities" as including "product and enterprise analysis; the value and importance of codes, interpretation and marketing, the potential significance of new community and political mechanisms; and post-trip lifestyle change."

Not covered by Weaver and Lawton is the well-established trend of explicit critiques of ecotourism, which have included scholars questioning whether it can promote sustainable development (Bramwell and Lane 2005), problematizing both definitional problems and issues of implementation (Bjork 2007), implicating it in "neoliberalizing nature" (Duffy 2008) and critiquing it as "egotourism"—a "Trojan horse" inevitably entrapping local economies toward mass tourism (Wheeler 1991: 96). Another classic (albeit outdated[3]) body of ecotourism critiques concerned itself with the commodification of culture (Silver 1993) and the problems arising from the "Westernization" of the locals (Cohen 1989) and acculturation to capitalist translations and "manufacture" of culture (Dearden 1991). What many critiques have in common is they focus on what ecotourism does or does not do, rather than the fact that it may be complicit in beyond itself. There are also scholars like Cater (2006: 23) who explicitly theorizes ecotourism as a Western construct, inscribed into and complicit in reproducing "Western cultural, economic, and political processes"; Carrier and Macleod (2005), who deconstruct the ecotourist "bubble" that abstracts the tourists from the places they purportedly come to experience; and Duffy (2002, 2010), who not only identifies ecotourism as one form of neoliberalization of nature, rather than an alternative to it, but who even makes an argument that ecotourism itself can be understood as an "extractive" industry.

Methodologies

One more aspect of ecotourism that Weaver and Lawton do not address specifically is the range of works that try to develop and articulate methodology for studying ecotourism. To some extent the methodological bias of all researchers are implicit in the research goals and frames in the profiled texts. However, before moving on to discussing extractive industries, I briefly review articles that explicitly propose or demonstrate methodologies of how to study ecotourism, and discuss the ways in which many of the methods are at odds with a structural, ecosystem-scale analysis that would flag the "paradoxes" (or note the convergences) of the ecotourism-extraction nexus.

Some authors who fall under this rubric discuss the methods of "valuing" ecotourism, such as the "travel cost method" (Tobias and Mendelsohn 1991) or engage in a cost-benefit analysis (Buckley 2009). Increasingly, in the last decade the "contingent valuation" method has been used (Lindsey et al. 2005), especially within the framework of ecological economics and its flagship journal, *Ecological Economics* (Baral et al. 2008; Lee and Mjelde 2006); Zografos and Allcroft (2007) developed a method for a market segmentation study based on the environmental values of potential ecotourists. Others, like Thomlinson and Getz (1996) offer a framework for research and evaluation of the contested issue of scale in ecotourism.

In general, there appears to be a trend toward segmentation in methodologies themselves— micromethodologies that frequently overlap in either goals or executions seem to place certain kinds of disciplinary "stamps" on the studies in question. Of course, from the discipline of anthropology, or at least from practitioners of ethnography there are many rich case study analyses that make site-specific methodological contributions (Chernela 2005b. 2011; Moreno 2005; Stronza 2005). These studies are all the more methodologically valuable because as a body of knowledge they provide a counterpoint to virtualism and essentialism of normative, anti-ethnographic frameworks of what ecotourism "is" and what it "does". Some case study analyses in this vein, like Fletcher's (2009) ethnographic perspective on ecotourism development in a community in southern Chile explicitly challenges the conventional methodologies of studying ecotourism in "stakeholders theory" approaches and emphasizing ecotourism as a discursive process deeply contingent on local, emic understandings of the world.

Critical Studies of Resource Extraction

Although ecotourism studies span a range of disciplines, there is a relatively established niche for subject-specific studies of ecotourism in both applied and critical social sciences. "Critical studies of resource extraction" is a less defined niche of production of knowledge, although plenty of knowledge production on the subject takes place. Ecotourism as a phenomenon was (arguably) "launched" institutionally within a framework that at least nominally invited reflexivity and assessment. However, resource extraction in and of itself is not a "designed alternative" industry, but an established mainstream capitalist practice, which has generated endless variations, oppositions, and critiques. It is not my intention to argue that ecotourism is somehow a less complex subject of study than resource extraction, nor that the ethnographic experiences with it are in any way more uniform that the complex engagements between communities and extractive industries worldwide. On the contrary, one of the challenges of any critical study of ecotourism is to show the multiplicity and the diversity of the agendas, agencies, and experiences that are not contained by the normative narrative of ecotourism as universally applicable solution that is "wholly benign, environmentally, culturally, and economically" (West and Carrier 2004: 484). But while both fields of study contend with multiple forms of knowledge and material practices, the "umbrella" concept of ecotourism is that it is an institutionally constructed framework that practices self-articulation and self-identification as a coherent entity much more than what we understand as "extraction." To put it another way, the epistemic community of practitioners and academics around ecotourism is diverse, decentralized, and by no means in agreement. Nevertheless, it overlaps enough that it has a shared vocabulary of ideals and practices, although some may try to implement those ideals and practices, while others may critique them from different perspectives. But there is no one unified, however broad, area of "expertise" around resource extraction, which as a field of practical knowledge has a culture of expertise that is far more scientist and technical than anything in the ecotourism sector. For that reason, the literature reviews of the two sectors of knowledge have to be organized in a different way, although some of the epistemological categories may be common to both. I focus on debates and trends in applied and critical literature around resource extraction—that is, works primarily by social scientists. As with the previous section, the literature review I offer in this article, while hopefully comprehensive or at least sufficiently representative, is informed by my agenda to identify theoretical and methodological trends that may impede the study of co-occurring processes of ecotourism and extraction—and, conversely, to recognize the trends that can enable such study.

Theoretical and Epistemological Issues in Resource Extraction Studies

As per Erich Zimmerman's (1933) famous quote, "resources are not: they become." First of all, it should be mentioned that there is an ever-growing field of literature that does the heavy theoretical lifting on the subject of how resources are constructed; how they become—these questions and processes belie studies that deal with the ascribed "nature" and "value" of extractable resources. Most recent are new resource geographies and ethnographies that focus on temporality (Ferry and Limbert 2008), and materiality (Bakker and Bridge 2006; Kaup 2008) of resources. In some cases even the agency of resources is theorized. For example, Orlove and Canton (2010) have proposed that water be regarded as a "total social fact" due to the social and symbolic possibilities encoded in water's materiality and its molecular properties, while Bakker (2004) titled her book about water *An Uncooperative Commodity*.

These theoretical perspectives are the vanguard of the established within political ecology ways of studying what Kaup (2008: 1735) describes as "new forms of organization within nature-based commodity sectors, which often function to secure, define, and limit the access to certain types of nature and their benefits." Scholars and works reviewed in this section contribute in different ways to the work on extractable resources and their social and political lives. They consider different interpretations of resource regimes from the controversial "resource curse" to "resource triumphalism" (Bridge 2001) and illuminate loci where issues of environmental governance, environmental change, and environmental valuations are played out as subjects of ongoing debates about costs and benefits, opportunities and vulnerabilities, exploitations and exchanges, enactments of agency and transfers of sovereignty, economics and "development."

Probably the most prominent and well-known theory (and theoretical debate) around resource extraction in the social sciences is that of the so-called resource curse. This theory, first articulated by Auty (1993) and widely used by Karl (1997) and Ross (1999) stipulates that resource-rich developing countries are "paradoxically" unable to mobilize their resource wealth for economic growth, and, over time, experience worse economic growth indicators and outcomes than countries that lack resource abundance. Studies like Sachs and Warner's (1995) cross-country comparison have argued for a negative association between resource abundance and economic growth. Scholars from different disciplines subsequently either endeavored to refine the theory, like Hodler (2006) who argued that natural resources only impeded growth in fractionalized countries, or tried to provide an explanation for the "curse," linking mineral wealth to the structure and quality of the political system (Jensen and Wantchekon 2004) or emergence of enclave economies or rentierism (Franke et al. 2009). Others have applied this theory to specific case studies, such as nickel, copper, and gold in Botswana (Iimi 2006), or oil in Chad and Cameroon (Pegg 2005). But there is also a body of literature that interrogates and critiques the entire framework of the resource curse and the epistemological assumptions that inform it. Thus, Watts (2004) critiques the theory as suffering from commodity-determinism and inadequate attention to the impact of specific resource characteristics on politics and conflict; while Lahiri-Datt (2006) draws on her work on informal mining practices to show how the normative hidden assumptions about resource ownership in the "resource curse" framework delegitimize the subsistence practices (and agency) of a wide range of communities. In turn, Weszkalnys (2009: 698) speculates that the "resource curse" literature may be an "academic fad of scholars seeking to engage in a form of 'useful science.'" However, she points out that it has usefully opened up a new research agenda "that seems to shake up the foundational categories of the disciplines of international relations, economics, and political sciences."

Do these epistemological paradigms and trends contribute to the blind spot of the ecotourism-extraction nexus? Can they be mobilized to study the nexus? The "resource curse" paradigm echoes certain discourses about ecotourism (such as the "curse or blessing?" question explicitly asked by the likes of Myles [2003] and the discussion of the "curse of success" in sustainable tourism by Stephan [2000]), and offers some resonances and parallels with the "dispossession by conservation" (Gordon 1989) or "eviction for conservation" (Brockington and Igoe 2006) phenomena. Yet it implicitly but powerfully a) demarcates "resources" as extractable and exportable, and b) is firmly rooted in macroeconomics and macropolitics, with its focus on state structures and economic growths. This perspective makes a locale-based or an ecosystem-based focus challenging for reasons that resonate with Lahiri-Datt's critique. However, if the open research space noted by Weszkalnys can help facilitate a form of analytic engagement with the concept of "resources" that is attuned to paradoxes, and is equally attentive to different types of nature "resources" as potential sources of "goods" or "value," regardless of their physical mobility or export possibilities, that may be a constructive research direction.

Different genres of critical engagement required for a nexus-centered approach is already present in much of existing scholarship on the political ecology and ethnography of resources and their extraction. Three seminal review articles published almost two decades apart reflected on the production of knowledge about mining, establishing emergent frameworks for "explor[ing] the nexus of the symbolic and the material" while grounded in "concrete locations" (Biersack 2006). Godoy's now-classic 1985 text "organized" the mining industry as something that could be studied within the frameworks of economic base, social organization, and ideology. Bridge's (2004) review focused on the intersection of mining, environment. Ballard and Banks's (2003) review focused less on thematic typologies and more on different positions and constructions of actors in anthropological studies of mining; their critique of anthropological tendencies "to maintain their focus on the more familiar 'exotic,' addressing the position of local communities in the vicinity of mines in preference over the less familiar multinational mining corporations", resulting in "the figure of 'the mining company' lurk[ing] monolithically and often menacingly in the background of many anthropological accounts of communities affected by mining operations" attested to the necessity of ethnographic engagement with the "the exceptional complexity of the relationships that coalesce around mining projects" (Ballard and Banks 2003: 287)—and other forms of resource extraction. Such complexities can (and do) include situations where ecotourism and extraction converge in the same locales, as they either oppose or enable each other.

As with ecotourism literature, I do not wish to recapitulate existing reviews, but to identify trends in the production of knowledge that help illuminate how disciplinary and subdisciplinary differences in "critical resource studies" are constructed and reproduced, and to look at which units of analysis and what sorts of scales tend to be used and privileged by different approaches. What kinds of researchers take community-level or ecosystem-level analyses as their basis? Do specific objects of inquiry and specific disciplinary grounding inspire more policy-level or state-level analysis?

Issues of Scale and Frame in Resource Extraction Studies

Some significant—although by no means exhaustive—trends in the studies of resource extraction include an increasing emphasis on the links between resource extraction and "empire building" on a global scale and the "petroviolence" (Watts 1999) that comes with this project, as well as state building through resource-led development. Reyna frames his 2011 piece on emergent "monster" beliefs in oil consortium "sorcerers" in Chad with a critical perspective on the importance of oil to imperial domination. Friedman (2010) examines the global networks of power around oil. In critiquing what he calls a "resource determinist approach" to oil and violence he attributes to Karl (1997) and Ross (1999), Friedman argues that it is not oil production itself that is important but "the way it is appropriated and re-deployed within the world economy" (2010: 32). In a complementary engagement of the global scope of oil, Ferguson (2005) follows up Scott's (1998) "seeing like a state" perspective by analyzing oil-related (and thusly territorialized) capital investment in Africa and the newly emergent forms of spatialized order and disorder. Hornborg (2009) draws on the world-systems theory to think through the ways in which the assumptions and expectations of the cultural context of the capitalist world-system construct peak oil. Zalik's "oil futures" text (2010) links discussions and perceptions of oil "scarcity" with the global social construction of oil prices.

It is not coincidental that all the macrolevel political economy analyses reviewed above connect to one particular form of extraction—oil extraction. In "extraction literature," oil has long been a site of exploration of issues of global economy and "empire" on global scales. Other forms of extraction, especially mining and logging, while certainly engaged through a political economy lens, feature more in studies of nation building and local nuances than global dominations—although there are some efforts in that direction (Luning's [2010] follow-up to Scott and Ferguson, "Seeing Africa Like a Mining Company" is one recent example). But mostly, when mining is considered "globally" it is listed among other forms of global exploitation (Bales [1999] names mining as one of the industries that uses slave labor in the "new global economy" along with prostitution, gemworking, and clothmaking), or other types of global multinational enterprises (Dunning and Lundan 2008), or other resource-intensive but not necessarily extractive sectors of global economy like construction, transport, and agriculture (Behrens et al. 2007).

This is not to say that oil extraction and, for example, mining, or other forms of resource extraction should always be analyzed within the same scales. One of my research sites is Ecuador, a nation where the current government has strategically juxtaposed oil and mining industries, using them to represent different stages in Ecuador's history of itself (with the oil sector symbolizing neoliberal "baggage" and mining standing for populist, state-controlled, resource-led "new" development). Based on my research in Intag, Ecuador—a place where anti-mining activists identify more with anti-mining activists from other countries than with fellow Ecuadorian anti-oil activists because, among other articulated reasons, "oil and mining are just different"—I am convinced that undifferentiated construction of different types of extraction as categorically similar "economic opportunities" or "environmental threats," while useful politically, may prove essentialist and anti-ethnographic. And there is certainly rich ethnographic work on oil extraction that is ethnographically embedded in local realities or in national narratives, even as it deals with the global politics of oil (Gilberthorpe 2007; Sawyer 2004). But there is something revealing about the fact that most recent explicitly theory-making works on oil (e.g., Reyna et al. 2011) draw on diverse local cases to theorize a global framework for studying oil, while Ballard and Banks (2003) encourage the epistemological pursuit of fragmentation, multiplicities, and fluid and complex relationships, in response to the knowledge practices characterized by the monadic dark specter of "the mining company."

This divergent directionality may be connected to certain realities of how the respective extractive sectors under discussion are organized. Thus, there is a big difference between large-scale mining by a transnational corporation and artisanal mining by locals, but there is no such thing as artisanal oil extraction (unless the oil in question is of the olive variety). It may also be the case that oil is more of an undifferentiated phenomenon in Western intellectual public consciousness. There are "oil wars" and "oil imperialism" and "petro-violence" but also "alternatives to oil." Although there is some work on "conventional" versus "unconventional" (like oil sands) oil sources, the distinctions among varieties of hydrocarbons, such as differences between extra-heavy, heavy, light crude, bitumen, or sweet versus sour crude, remain more esoteric in social sciences and in the public consciousness than the difference between, for instance, copper and uranium, or aluminium and sapphires. That may have something to do with why academic work on oil easily lends itself to productive synergy with neo-Marxist political economy that is structured through a study of discrete categories like primary accumulation and added value. At the same time discourses about "mining" (of different kinds of minerals like copper, aluminium, bauxite, uranium, as well as rare earths, and precious and semi-precious stones) are, so far, less oriented to a macropolitical economy analysis, and more attuned either to local and national contexts (as evidenced by the numerous mining-related examples in the next section) or to locally grounded explorations of how transnational commodity chains of specific resources are connected to specific sectors of the global economy (see, for example, Mantz's [2008] analysis of the way coltan production in Eastern Congo services the digital demands of the global economy of knowledge, or Tsing's [2003] analysis of the interplay between the local and the global through the lens of the logging boom in Indonesia and the commodity chains of timber consumption). In addition, and relevantly for considering extraction and ecotourism in tandem, certain forms of mining—but never oil extraction—work with an interesting subset of resources that can be thought of as "amalgam objects" that, as commodities, in their constitution, representation, and interpretation blur the line between resource extraction and ecotourism. These are objects, which, through their particular "uniqueness" (ascribed or constructed though it may be) mediate the paradoxes between the categories of capitalism and nature, or extraction and ecotourism even as they reproduce them. The most obvious examples of these types of "amalgam" or "crossover" objects are certain types of gemstones, symbolically linked to specific locations. As demonstrated by Walsh's (2012) work on how Malagasy sapphires are infused with place-contingent value, and are fetishized in the same way that Malagasy nature is fetishized, or (illustratively) by a recent report titled "The Perfect Setting: Diamond Tourism in the Northwest Territories" prepared by a consulting firm[4] to facilitate diamond tourism in the Canadian North and linking and advocating capitalizing on the "perfect marriage combining the unique experiences of northern travel and the magic of the aurora borealis, with the purity and beauty of Canadian diamonds." Perhaps such phenomena are a part of why mining studies tend to be scaled locally, while oil is constructed as a global resource with local impacts.

In the discussion of scales in the production of knowledge about resource extraction, a backlash of sorts among a certain class of extraction scholars should be noted—a reaction to what Le Billon (2008), in his analysis of "diamond wars" calls "[the dominance] of econometric approaches and rational theory interpretations" that oversimplify or overlook the geographical dimensions of contested resources. Even without "rational theory interpretations," Marxist analyses like Labban's 2008 "Space, Oil, and Capital," which takes a global scale approach to the relationship between the production of oil, capitalist competition, and social production of space, start with large-scale economic categories and use them to illuminate and explain resource geographies as products of geopolitical practices. The more "grounded" approach of much of mining literature starts with particular geographic and ethnographic loci, producing a different kind of analysis.

Having noted these general trends that, for the time being, can be said to distinguish anthropology and political ecology of oil from anthropology and political ecology of other forms of extraction, what follows are some thoughts on significant trends in approaches and frameworks in critical resource extraction studies, across different types of resources. Certainly, the categories are nonexhaustive, and in many cases the texts discussed fit into more than one category. Thus, the reader should approach this not as a rigid taxonomy, but, rather, a reflection on coalescent topic clusters.

Social Movements and Mobilizations

There is an entire corpus of ethnographically rich and theoretically innovative literature that focuses on social movements and political mobilizations around resource extraction: among them Sawyer's (2004) work on the Ecuadorian oil patch; Kirsch's (2007) work on the indigenous resistance to the Ok Tedi mine in Papua New Guinea, with a particular focus on the dangers for indigenous actors of deviating from an "antidevelopment" position "expected" of them; Dougherty's (2011) take on community resistance to gold mining by Canadian junior firms in Guatemala and, also Guatemala-focused, Urkidi's complementary (2011) reclamation of a "community" as a legitimate unit of decision making in mining conflicts; Ali and Grewal's (2006) and Horowitz's (2002) work on community responses to nickel mining in New Caledonia; Kuecker's (2007) analyses of the emergence of environmentalism as a response to a copper mining threat in Ecuadorian highlands; Watts's (2008) take on the militancy and "petroinsurgency" in the Niger Delta; Welker's (2009) case study of village leaders and rural youth in Indonesia living near a mining company's operation attacking visiting environmental activists; Kaup's (2008) article detailing how Bolivia's social movements have approached and utilized natural gas to strategically engage with the processes of capital accumulation. The various theoretical concepts developed out of these case studies include "folk environmentalism" (Kuecker 2007), "environmentalism of the poor" (Guha and Martinez-Alier 1997), and "glocal environmental movement" (Urkidi 2011). Many of these texts, in one way or another, engage with Agrawal's notions of environmentality—a key concept in work on environment-and-resource related social movements (although no less key is Cepek's [2011] critique of Agrawal and his meditations on the limits of "environmentality" when it comes to emic categories of indigenous ethnoecologies).

In some cases, ecotourism (and other strategies of constructing nature as a site of inalienable value, rather than circulating goods), is part of the strategic opposition to resource extraction (Kuecker 2007; Stem at al. 2003) and is noted as such in case studies, but often such a perspective reinforces the purported dichotomy between the two, even when merely documenting the ethnographically relevant uptake of this dichotomy for local political uses. Thus, the community-scale focus dominant in "social movements" literature is a helpful platform for exploring the ecotourism-extraction nexus where it occurs, with the caveat that it is more likely to "catch" the cooccurances in an oppositional dynamic. When the two phenomena are in synergy, that kind of arrangement is unlikely to be on the radar of a "social movements" approach. This is because while increasingly there is work that destabilizes the simplistic David versus Goliath scenario of "community" versus "corporation," the social movement framework generally focuses on specific, recognizable types of social mobilizations or organized activity that makes tensions and ruptures explicit and visible in concrete ways. Arguably, instances of dispossession for conservation projects and ecotourism produce different (and perhaps understudied) social movements than anti-extraction mobilizations, not least because the alliances and types of access to transnational resources of media and political capital that are possible for anti-extraction activists are often denied to actors dispossessed for "green" reasons.

"Impactology": Extraction

As with ecotourism, there is a broad category of critical resource extraction studies that could be characterized as "impactology," where the primary goal of the research is to study the impacts on a particular set of actors or subjects. Because the range of effects and subjects is extremely wide, this may seem like an ineffective category to posit, but it does highlight a particular cluster of research practices and goals—usually concerned with changes over a designated period of time, and focused on specific social actors, or specific environments, whether or not the final goal is to use those actors or environments to represent a larger class of actors and environments.

This is the category that lends itself well to both quantitative and qualitative research rubrics, and is thus as likely to draw economists and development studies scholars or public health scholars as it is ethnographers and other more qualitatively oriented researchers. Lockie et al.'s (2009) longitudinal social impact assessment of the Coppabella coal mine on the community using the "resource community cycle" model; public health assessments like the study of para-occupational and environmental mercury exposure in small-scale gold miners in Nicaragua (Cuadra et al. 2009); a discussion of community health issues during both the "boom" and the "bust" parts of the mining cycle in a northern Canadian mining community (Shandro et al. 2011) represent one end of the research spectrum possible within the "impactology" parameters. At the other end are holistic ethnographic explorations of direct and indirect consequences of extractive industries on various domains of community life—from Walsh's (2003) discussion of "hot money" and "daring consumption" by young men involved in sapphire mining in Malagasy to the effects of oil money on kin relations in farmer communities of Southern Chad (Hoinathy 2011). Many of these works focus not so much on "material" changes, but on changes in beliefs and ways of meaning-making connected to what often amount to new moral economies.

In general, starting with modern "classics" like Taussig's (1987) writings on indigenous beliefs emerging around the terror of rubber plantations on the Putumayo River, a rich body of work documenting and theorizing both "occult economies" and broader religious and cosmological transformations around resource extraction has emerged: Werthmann (2003) on beliefs about ill-gotten gold-mining wealth in Burkina Faso; High (2008) on cosmological mediations of envy around mining wealth in Mongolia; Ekholm Friedman (2011) on child witchcraft during the "oil boom" of Congo-Brazzaville; Horowitz (2008) on the way Christian discourses frame local interpretations of multinational mining in New Caledonia; Golub (2006) on the fundamental disharmonies between Ipili cosmologies and the labor regime of the mining industry. At times these writings overlap with the more general work on the occult economies of neoliberalization, development, and accumulation (Comaroff and Comaroff 2001; Geschiere 1997) as well as with scholarship that looks at emergence of resources as contemporary forms of "total social facts" stressing the commodity fetishism around extraction practices (Nash 1979; Taussig 1980; Walsh 2010).

As with ecotourism, a certain type of "impactology" research into resources extraction can at times be limited by its own rubrics, resulting in a tunnel-vision research design that does not place chosen empirical metrics into broader structural context. But as a part of a larger structural analysis in a research framework characterized by a mixed-method approach, the emphasis on documenting changes over time, and the predilection for community-level or eco-system-level, "impactology" can be useful for empirical assessment of complex situations such as the copresence of extraction and ecotourism. In fact, the feature that most "impactology" studies, across the disciplines, have in common is the sort of community-level analysis that provides a good platform for a nexus-cognizant research orientation. Community-level analyses can allow dynamic studies that show how resource extraction industries not only impact specific aspects of community life, but also coexist and interact with other local industries and realities. Although not dealing with ecotourism per se, in terms of setup and methodology, studies like Cartier and Burge's (2011) work on synergies between farming and mining cycles in Sierra Leone, which argues that small-scale agriculture and artisanal mining are livelihood complements, rather than alternatives, exhibit the kind of dynamic study set-up and methodology that could facilitate a fruitful study of the ecotourism/extraction nexus.

Furthermore, the discussed "occult" subsection of the impactology literature is particularly interesting, as transformations of beliefs can illuminate how tensions, contradictions, and paradoxes emerge—local beliefs do not necessarily distinguish between categories like "sustainable" or "mainstream" development, and, in territories where both types of "development projects" arise, can point to emic perspectives that interpret both as new, and not necessarily different, types of "modern" incursions. These perspectives may reflect true ethnographic experiences of the locals. In my fieldsite in northern Russia, rapid privatization of nature in the post-Soviet period has transformed the beliefs of the Veps, a small indigenous group in a way that makes it visible how both resource extraction and ecotourism projects have become incorporated into their cosmology and discourse as similar, rather than different forces. Whereas historically their cosmology has pivoted around balanced and fair exchange with spirit masters of forests and lakes, their cosmological narratives have recently shifted so that these spirit masters are imagined to have departed due to being supplanted by the new "bad" masters. The latter are private companies who are both aggressively amplifying the logging and the mining activities in the region, and developing the Veps' lakeshore for luxurious eco-resorts, literally extracting portions of the beaches from local use and passage, both experienced as negative forces of privatization. In my other fieldsite, in the lowland Kichwa villages in Ecuador, narratives circulate of both resource extraction and ecotourism making appearances in shamanic visions (Davidov 2008; Whitten and Whitten 2007). Although the visions take place at different points in time, such "sightings" are often discussed together, as illustrations of the power of shamanism to foresee upcoming significant changes. Not only do extraction and ecotourism coexist in the lowlands of Ecuador, they explicitly converge as prophesied forms of "modernity" for communities becoming entangled with them.

Discussion

When Ballard and Banks (2003) wrote their review article assessing the state of knowledge production of the field, they were explicit about the fact that they felt compelled to do so in response to the "mining boom" as well as the increased recognition of indigenous rights from the 1980s on, and the institutionalization of impact assessments and community inclusion, thus highlighting a much larger pool of actors with complex relationships with mining companies, NGOs, and the state. This research dynamic, focused on a larger pool of social actors is visible in social science literature on human-nature relationships in general. Overall there has been a trend in ethnographic studies of both on-the-ground and transnational dynamics around extraction to follow Ballard and Banks's call, as monolithic and essentialist representations of local communities engaged in struggles over resources give way to "[drawing] upon historically sedimented practices, landscapes, and repertoires of meaning … through particular patterns of engagement and struggle" (2003: 98).

This practice in scholarship offers a counterpart to the literature that theorizes and narrates contested resource extraction (especially on indigenous territories) as marking the shift from "traditional" to "neoliberal" economic and social forms. One of the common narratives about neoliberalizaion of natural resources focuses on the rapid advent of extractive industries in previously "untouched" or "remote" areas (Bebbington and Bebbington 2010; Kuecker 2007; Tsing 2003) resulting in dispossession of the locals. The extent to which such narratives are ethnographically "true" is epistemologically ambiguous. Claims and articulations of "traditional" forms of subsistence are themselves contingent on historical positionalities formed during a succession of "modernities." Particularly in cases of indigenous communities, who still have to negotiate what Prins (1997) described as "paradox of primitivism," a discursive distinction between "tradition" and "neoliberalism" can be politically advantageous (Robins 2003). The advent of resource extraction often becomes an iconic moment in discursively separating regimes of "ecological integrity" (and "traditional" subsistence forms) from regimes of "ecological disembedding" (and extractive capitalism). In the same vein, strategic use of environmentalist rhetoric can, in some cases, link the advent of ventures like ecotourism with a reclamation of territories, and either opposition to or reversal of extractive activities. Since such "oppositional" discourse is frequently profiled by the media, and is emphasized by the pedagogical scripts of many ecotours, this becomes the mainstream understanding of what extraction and ecotourism can do for and to communities, and how they relate to each other. Given that trend, increased anthropological and critical geography engagements with cases where the inception of resource extraction is not and cannot be framed as the iconic moment of dispossession, the line in the sand between ecological holism and alienation, as well as the cases where ecotourism is not an "empowering" activity but a site of dispossession by conservation (Brockington and Igoe 2006), or an instance of corporate greenwashing, make a valuable contribution to anti-essentialist political ecology (Escobar 1999) and the critical study of human-nature relations. This non-essentialist ethnographic approach to extraction and to ecotourism that highlights the plurality of meanings and stakes involved, exemplified by work like Slater 2002; Walsh 2003; Carrier and Macleod 2005; Wadley and Eilenburg 2005; Stonich 2006; Vivanco 2006; and West 2006b can only be enhanced and thickened by empirically engaging with the nexus of ecotourism and extraction, as more and more fieldsites lend themselves to that kind of study.

So far, though, it appears that no research design exists that explicitly theorizes them as a nexus. There are many books, both monographs and anthologies that critically engage with ecotourism in ways that range from concept critiques to "impactology." There are also sophisticated examinations and critiques of resource extraction within frameworks of environmental anthropology, critical geography, rural sociology, and other disciplines that draw on the political ecology approach. But, on the whole, even sophisticated analyses of ecotourism and resource extraction either explicitly (through juxtaposition) or implicitly (through omission) engage these two concepts as divergent development trajectories. This is the case even when the field and the study scope lend themselves to a "nexus" perspective, although there is some existing work that may prefigure such an approach (e.g., Che's [2006] work on the efforts of Forest County in Pennsylvania to diversify its economy by developing ecotourism based on its unique hardwood forests produced by timber harvesting).

Once a theoretical foundation for studying ecotourism and extraction together is in place, and once it is recognized as a necessary and desirable way of focusing and framing research, what are the practical ways in which the two phenomena can be studied as a nexus? Can there be an empirical model, a methodological approach that would be applicable in multiple fieldsites? I conclude this article by proposing two such approaches, in hopes that many more will be generated by scholars working in "nexus" fieldsites currently and in the future.

One approach could be based in empirical, ecosystem-scale or locality-scape analysis of a place where ecotourism and resource extraction are copresent. It could be achieved through environmental ethnographies that draw comparative ethnographic and ecological analysis of communities in "nexus" areas and focus on the range of experiences and outcomes for local actors, empirically engaging with the common assumptions that resource extraction degrades the environment and harms its denizens, while ecotourism conserves the environment and benefits its denizens (and that, consequently, ecotourism can be a part of the "environmental offset" for resource extraction, the way it is supposed to be in the Chad-Cameroon pipeline case). The optimal way of studying the nexus through this approach would be a combination of ethnographic projects and ecological surveys designed to empirically assess the impact of ecotourism and resource extraction on the communities experiencing the effects of both. The ecological surveys would need to overlap in time with ethnographic fieldwork to ensure mutual dialectical feedback, where, for example, discoveries of certain environmental impacts could then be addressed ethnographically. Conversely, ethnographic research could yield data that may suggest environmental changes that would need to be documented and analyzed. Of course, all ethnographic and ecological data would be contextualized by a historical and institutional analysis of the central actors, networks, and dynamics pertinent to the political ecology of the region.

The second approach focuses on the aforementioned phenomena that blur the line between resource extraction and ecotourism. Such "crossover" objects—often types of stones or wood—are promoted in identification with the nature of the places. They are products of natural processes,[5] and valued as such, yet in order to be marketed and sold they have to be extracted. So, such "amalgam," or more accurately, "integrative"[6] objects are fundamental to both extraction and ecotourism enterprises in the same locales, and thus can be starting points of study and key figures in interpretive frameworks of how nature is constructed, valued, and used in these places. In addition, activities like "trophy hunting" in safari tourism can be interpreted in the same way (as explicitly extractive activities which are, nevertheless, iconic aspects of ecotourism in that particular location). Sometimes, iconic activities (rather than objects) associated with the two respective industries themselves merge into a single amalgam, as in cases where versions of extractive activities become a part of the "ecotour package" (so, panning for gold is a staple part of ecotours in Ecuadorian Amazon), or in cases of "recreational mining." "Crossover" phenomena like that can serve as "theory machines" (Galison 2003; Helmreich 2011) to reconceptualize the schismogenesis between resource extraction and ecotourism and help destabilize the categorial difference between the two on a conceptual level. This approach, although inherently limited, can lead to either theoretical reconceptualizations of how a specific site of nature is being constructed and "developed," or can serve as an entry point that uses a "materiality" frame for an environmental ethnography grounded in a political ecology perspective.

Both approaches proposed in this coda can build on much of the literature reviewed in this article, and achieve a new and warranted research agenda: a systematic (and, ideally, comparative) study of the copresence of resource extraction and ecotourism, including how the range of possibilities and reasons for why such a convergence may manifest, and the range of outcomes of this copresence for the local actors and landscapes.

■ ACKNOWLEDGEMENTS

I am grateful to Paige West and Dan Brockington for providing conceptual and editorial feedback during the writing of this article. I also thank the three anonymous reviewers for the feedback on the first draft of this text; Bram Büscher for helping me integrate the reviewers' suggestions and for providing additional feedback; the participants in the 2012 European Association for Social Anthropology panel on extraction and ecotourim, who inspired me with their diverse range of ethnographic case studies; and Barabara Andersen for advice on and help with structure and coherence.

■

VERONICA DAVIDOV is assistant professor of Anthropology at Leiden University College. Her research interests include the production of normative and contested discourses of nature and human-nature relations, the transformation of nature into natural resources, the impact of globalization and "development" (including "sustainable development") on indigenous cultures, and indigenous ethnoecology. She has done long-term fieldwork in Ecuador since 2002, and is currently also working on a project in northern Russia.

■ NOTES

1. Abstracts submitted for a panel I coorganized for the 2012 European Association for Social Anthropologists meeting on the topic included further ethnographic examples of such copresences: between rubber plantations and ecotourism in Laos; larimar mining and ecotourism in southwestern Dominican Republic; the cut flower industry and ecotourism in Kenya; ecotourism and oil extraction around the Sartsoon-Temash National Park in Belize.
2. Anonymous review for the Integrated Programmes 2010 WOTRO subsidy, project # W 01.67 .2010.026.
3. Over time the "hosts" and "guests" dichotomy approach morphed into ethnographically thick explorations of the nuanced dynamics of the performances (on both sides), spaces and encounters of ecotourism.
4. The North Group, "The Perfect Setting: Diamond Tourism in Northwest Territories" (2004), http://www.iti.gov.nt.ca/Publications/2007/Diamonds/diamond_setting.pdf (accessed 1 April 2012).
5. Although crude oil and uranium are also natural substances, they do not represent the "pure" "authentic" nature in a way that can be consumed and marketed.
6. "Integrative" can describe the function of such "amalgam" objects, because amalgamation pertains to how the objects have to straddle two industries; but as "theory machines" these objects have an integrative function as centerpieces of sorts in the imaginary Venn diagrams between extraction and ecotoursim in specific places.

■ REFERENCES

Agrawal, Arun. 2005. *Environmentality: Technologies of Government and the Making of Subjects.* Durham, NC: Duke University Press.

Ali, Saleem H., and Andrew S Grewal. 2006. "The Ecology and Economy of Indigenous Resistance: Divergent Perspectives on Mining in Caledornia." *Contemporary Pacific* 38, no. 2: 293–325.

Auty, Rochard. 1993. *Sustaining Development in Mineral Economies: The Resource Curse Thesis.* New York: Routledge

Bakker, Karen. 2004. *An Uncooperative Commodity: Privatizing Water in England and Wales.* Oxford: Oxford University Press.

Bakker, Karen, and Gavin Bridge. 2006. "Material Worlds? Resource Geographies and the 'Matter of Nature.'" *Progress in Human Geography* 30, no. 1: 5–27.

Bales, Kevin. 1999. *Disposable People: New Slavery in the Global Economy.* Berkeley: University of California Press.

Ballard, Chris, and Glenn Banks. 2003. "Resource Wars: The Anthropology of Mining." *Annual Review of Anthropology* 32: 287–313.

Baral, Nabin, Marc J. Stern, and Ranju Bhattarai 2008. "Contingent Valuation of Ecotourism in Annapurna Conservation area, Nepal: Implications for Sustainable Park Finance and Local Development." *Ecological Economics* 66, nos. 2–3: 218–227.

Bebbington, Anthony, and Denise Humphrey Bebbington. 2010. "An Andean Avatar: Post-Neoliberal and Neoliberal Strategies for Promoting Extractive Industries." Brooks World Poverty Institute Paper 17, Manchester.

Behrens, Arno, Stefan Giljum, Jan Kovanda, and Samuel Niza. 2007. "Countries, Regions and Trade: On the Welfare Impacts of Economic Integration." *European Economic Review* 51, no. 5: 1277–1301.

Belsky, Jill. 1999. "Misrepresenting Communities: The Politics of Community-Based Rural Ecotourism in Gales Point Manatee, Belize." *Rural Sociology* 64, no. 4: 641–666.

Biersack, Aletta. 2006. "Reimagining Political Ecology: Culture/Power/History/Nature." Pp. 3–40 in *Reimagining Political Ecology,* ed. A.Biersack and J. Greenberg. Durham: Duke University Press.

Bjork, Peter. 2007. "Definition Paradoxes: From Concept to Definition." Pp. 23–46 in *Ecotourism: Understanding a Complex Tourism Phenomenon,* ed. J. Hingham. Amsterdam: Elsevier.

Blamey, Russell. 1997. "Ecotourism: The Search for an Operational Definition." *Journal of Sustainable Tourism* 5, no. 2: 109–130.

Blamey, Russell. 2001. "Principles of Ecotourism." Pp. 5–22 in *Encyclopedia of Ecotourism,* ed. D. Weaver. Wallingford, UK: CAB International.

Bramwell, Bill, and Bernard Lane. 2005. "Sustainable Tourism Research and the Importance of Societal and Social Science Trends." *Journal of Sustainable Tourism* 1, no. 1: 1–3.

Bridge Gavin. 2001. "Resource Triumphalism: Postindustrial Narratives of Primary Commodity Production." *Environment and Planning A* 33, no. 12: 2149–2173.

Bridge, Gavin. 2004. "Contested Terrain: Mining and the Environment." *Annual Review of Environment and Resources* 29: 205–259.

Brockington, Daniel, and Rosemary Duffy. 2010. "Capitalism and Conservation: The Production and Reproduction of Biodiversity Conservation." *Antipode* 42, no. 3: 469–484.

Brockington, Daniel, and Igoe, James. 2006. "Eviction for Conservation: A Global Overview." *Conservation and Society* 4, no. 3: 424–470.

Brockington, Daniel, Rosemary Duffy, and James Igoe. 2008. *Nature Unbound: Conservation, Capitalism, and the Future of Protected Areas.* London: Earthscan.

Brondo, Keri, and Natalie Bown. 2011. "Neoliberal Conservation, Garifuna Territorial Rights and Resource Management in the Cayos Cochinos Marine Protected Area." *Conservation and Society* 9, no. 2: 91–105.

Bruner, Edward M. 2005. *Culture on Tour: Ethnographies of Travel.* Chicago: University of Chicago Press.

Buckley, Ralph. 2009. "Evaluating the Net Effects of Ecotourism on the Environment: A Framework, First Assessment and Future Research." *Journal of Sustainable Tourism* 17, no. 6: 643–672.

Büscher, Bram. 2010. "Anti-Politics as Political Strategy: Neoliberalism and Transfrontier Conservation in Southern Africa." *Development and Change* 41, no. 1: 29–51.

Carrier, James, and Daniel Miller. 1998. *Virtualism: A New Political Economy.* Cambridge: Berg.

Carrier, James, and Donald Macleod. 2005. "Bursting the Bubble: The Socio-Cultural Context of Ecotourism." *Journal of the Royal Anthropological Institute* 11, no. 2: 315–334

Cartier, Laurent, and Michael Burge. 2011. "Agruculture and Artisanal Gold Mining in Sierra Leone: Alternatives or Complements?" *Journal of International Development* 23, no. 8: 1080–1099.

Castree, Noel. 2008. "Neoliberalizing Nature: The Logics of Deregulation and Reregulation." *Environment and Planning A* 40, no. 1: 131–152.

Cater, Erlet. 2006. "Ecotourism as a Western Construct." *Journal of Ecotourism* 5, no. 1&2: 23–39.

Cepek, Michael. 2011. "Foucault in the Forest: Questioning Environmentality in Amazonia." *American Ethnologist* 38, no. 3: 501–515.

Che, Deborah. 2006. "Developing Ecotourism in First World, Resource-Dependent Areas." Geoforum 37, no. 2: 212–226.

Chernela, Janet. 2005a. "The Art of Listening: Collaboration Between International Environmental NGOs and Indigenous Peoples in the Amazon Basin of Brazil." *Worldwatch* (February).

Chernela, Janet. 2005b. "The Politics of Mediation: Local-Global Interactions in the Central Amazon of Brazil." *American Anthropologist* 107, no. 4: 620–631.

Chernela. Janet. 2011. "Barriers Natural and Unnatural: Islamiento as a Central Metaphor in Kuna Ecotourism." *Bulletin of Latin American Research* 30, no. 1: 35–49.

Cheong, So-Min, and Marc L. Miller. 2000. "Power and Tourism: A Foucauldian Observation." *Annals of Tourism Research* 27, no. 2: 371–390.

Cohen, Erik. 1989. "Alternative Tourism—A Critique." Pp. 127–143 in *Towards Appropriate Tourism: The Case of Developing Countries,* ed. T. Singh, L. Theuns, and F. Go. Frankfort: Peter Lang.

Comaroff, Jean, and John Comaroff. 2001. "Millenial Capitalism and the Culture of Neoliberalism." Pp. 177–189 in *Anthropology of Development,* ed. M. Edelman and A. Hagerud. Oxford: Blackwell.

Corson, Catherine. 2010. "Shifting Environmental Governance in a Neoliberal World: US AID for Conservation." *Antipode* 42, no. 3: 576–602.

Cuadra, Steven, Thomas Lundh, and Kristina Jakobsson. 2009. "Paraoccupational and Environmental Mercury Exposure Due to Small Scale Gold Mining in Central Nicaragua: A Cross-Sectional Assessment of Blood Mercury Levels in Children and Women." *Epidemiology* 20, no. 6: S227.

Davidov, Veronica. 2008. "Shamans and Shams: The Cultural Effects of Tourism in Ecuador." *Journal of Latin American and Caribbean Anthropology* 15, no. 2: 387–410.

Dearden, Philip. 1991. "Tourism and Sustainable Development in Northern Thailand." *Geographical Review* 81, no. 4: 400–413.

Dougherty, Michael L. 2011. "The Global Gold Mining Industry, Junior Firms, and Civil Society Resistance in Guatemala." *Bulletin of Latin American Research* 30, no. 4: 403–418.

Duffy, Rosaleen. 2002. *A Trip Too Far: Ecotourism, Politics, and Exploitation.* London: Earthscan.

Duffy, Rosaleen. 2008. "Neoliberalising Nature: Global Networks and Ecotourism Development in Madagascar." *Journal of Sustainable Tourism* 16, no. 3: 327–344.

Duffy, Rosaleen. 2010. *Nature Crime. How We're Getting Conservation Wrong.* New Haven: Yale University Press.

Dunning, John, and Sarianna Lundan. 2008. *Multinational Enterprises and the Global Economy.* 2nd ed. Cheltenham, UK: Edward Elgar.

Eagles, Paul, and Cascagnette, Joseph. 1995. "Canadian Ecotourists: Who Are They?" *Tourism Recreation Research* 20, no. 1: 22–28.

Ekholm Friedman, Kajsa. 2011. "Elfs and Witches: Oil Cleptocrats and the Destruction of Social Order in Congo-Brazzaville." Pp. 107–131 in *Crude Domination: An Anthropology of Oil,* ed. S. Reyna, A. Behrends, and G. Schlee Oxford: Berghahn Books.

Escobar, Arturo. 1999. "After Nature: Steps to an Anti-Essentialist Political Ecology." *Current Anthropology* 40, no. 1: 1–30.

Fennell, David. 2001. "A Content Analysis of Ecotourism Definitions." *Current Issues in Tourism* 4, no. 5: 403–421.

Ferry, Elizabeth and Mandana Limbert. 2008. "Introduction." Pp. 3–23 in *Timely Assets: The Politics of Resources and Their Temporalities,* ed. E. Ferry and M. Limbert. Santa Fe, NM: School for Advanced Research Press.

Ferguson, James. 2005. "Seeing Like an Oil Company: Space, Security, and Global Capital in Neoliberal Africa." *American Anthropologist* 107, no. 3: 377–382.

Fletcher, Robert. 2009. "Ecotourism Discourse: Challenging the Stakeholders Theory." *Journal of Ecotourism* 8, no. 3: 269–285.

Franke, Anja, Andrea Gawrich, and Gurban Alakbarov. 2009. "Kazakhstan and Azerbaijan as Post-Soviet Rentier States: Resource Incomes and Autocracy as a Double 'Curse' in Post-Soviet Regimes." *Europe-Asia Studies* 61, no. 2: 109–140.

Friedman, Jonathan. 2011. "Oiling the Race to the Bottom." Pp. 30–45 in *Crude Domination: An Anthropology of Oil,* ed. S. Reyna, A. Behrends and G. Schlee Oxford: Berghahn Books.

Galison, Peter. 2003. *Einstein's Clocks, Poincare Maps: Empires of Time.* New York: W.W. Norton.

Geschiere, Peter. 1997. *The Modernity of Witchcraft: Politics and the Occult in Postcolonial Africa.* Charlottesville: University of Virginia Press.

Gilberthorpe, Emma. 2007. "Fasu Solidarity: A Case Study of Kin Networks, Land Tenure and Oil Extraction in Kutubu, Papua New Guinea." *American Anthropologist* 109, no. 1: 109–119.

Golub, Alex. 2006. "Who Is the 'Original Affluent Society'? Ipili 'Predatory Expansion' and the Porgera Gold Mine, Papua New Guinea." *Contemporary Pacific* 18, no. 2: 265–292.

Gordon, Robert. 1989. "Can Namibian San Stop Dispossession of Their Land?" Pp. 138–154 in *We Are Here: Politics of Aboriginal Land Tenure,* ed. E. Wilmsen. Berkeley: University of California Press.

Godoy, Ricardo. 1985. "Mining: Anthropological Perspectives." *Annual Review of Anthropology* 14: 199–217.

Guha, Ramachandra, and Juan Martínez-Alier. 1997. *Varieties of Environmentalism: Essays North and South.* London: Earthscan.

Helmreich, Stefan. 2011. "Nature/Culture/Seawater." *American Anthropologist* 113, no. 1: 132–144.

Heynen, Nick, James McCarthy, Scott Prudham, and Paul Robbins. 2007. *Neoliberal Environments: False Promises and Unnatural Consequences.* London: Routledge.

High, Mette. 2008. "Wealth and Envy in the Mongolian Gold Mines." *Cambridge Anthropology* 27, no. 3: 1–18.

Hodler, Roland. 2006. "The Curse of Natural Resources in Fractionalized Countries." *European Economic Review* 50, no. 6: 1367–1386.

Hoinathy, Remadji. 2011. "Monetization of Social interrelations in the Chadian Oil Zone: Money in the Core of Marriage, Kin Ties and Alliances." Paper presented at the American Anthropological Association Annual Meeting, Montreal, Canada.

Honey, Martha. 2002. *Ecotourism and Certification: Setting Standards in Practice.* Washington DC: Island Press.

Hornborg, Alf. 2009. "Zero-Sum World: Challenges in Conceptualizing Environmental Load Displacement and Ecologically Unequal Exchange in the World-System." *International Journal of Comparative Sociology* 50, nos. 3–4: 237–262.

Horowitz, Leah. 2002. "Daily, Immediate Conflicts: An analysis of Villagers' Arguments about a Multinational Nickel Mining Project in New Caledonia." *Oceania* 73, no. 1: 35–55.

Horowitz, Leah. 2008. "Destroying God's Creation or Using What He provided? Cultural Models of a Mining Project in New Caledonia." *Human Organization* 67, no. 3: 292–306.

Hughes, Karen, Jan Packer, and Roy Ballantyne. 2011. "Using Post-Visit Action Resources to Support Family Conservation Learning Following a Wildlife Tourism Experience." *Environmental Education Research* 17, no. 3: 307–328.

Hutchins, Frank. 2007. "Footprints in the Forest: Ecotourism and Altered Meanings in Ecuador's Upper Amazon." *Journal of Latin American and Caribbean Anthropology* 12, no. 1: 75–103.

Igoe, James. 2003. *Conservation and Globalization: A Study of National Parks and Indigenous Communities from East Africa to South Dakota.* Riverside CA: Wadsworth.

Igoe, James, and Dan Brockington. 2007. "Neoliberal Conservation: A Brief Introduction." *Conservation and Society* 5, no. 4: 432–449.

Iimi, Atsushi. 2006. "Did Botswana Escape from the Resource Curse?" International Monetary Fund Working Paper.WP/06/138.

Jensen, Nathan and Leonard Wantchekon. 2004. "Resource Wealth and Political Regimes in Africa." Comparative Political Studies 37, no. 7: 816–841.

Karl, Terry. 1997. *The Paradox of Plenty.* Berkeley: University of California Press.

Kaup, Brent. 2008. "Negotiating Through Nature: The Resistant Materiality and Materiality of Resistance in Bolivia's Natural Gas Sector." *Geoforum* 39, no. 5: 1734–1742.

Kirsch, Stuart. 2007. "Indigenous Movements and the Risks of Counterglobalization: Tracking the Campaign Against Papua New Guinea's Ok Tedi Mine." *American Ethnologist* 34, no. 2: 303–321.

Kuecker, Glen. 2007. "Fighting for the Forests: Grassroots Resistance to Mining in Northern Ecuador." *Latin American Perspectives* 153, no. 34: 94–107.

Labban, Mazen. 2008. *Space, Oil and Capital.* London: Routledge.

Lahiri-Datt, Kuntala. 2006. "'May God Give Us Chaos, So That We Can Plunder': A Critique of 'Resource Curse' and Conflict Theories." *Development* 49, no. 3: 14–21.

Laverack, Glenn, and Thangphet, Sopon. 2009. "Building Community Capacity for Locally Managed Ecotourism in Northern Thailand." *Community Development Journal* 44, no. 2: 172–185.

Le Billon, Philippe. 2008. "Diamond Wars? Conflict Diamonds and Geographies of Resource Wars." *Annals of the Association of American Geographers* 98, no. 2: 345–372.

Lee, Chong-Ki, and James W Mjelde. 2007. "Valuation of Ecotourism Resources Using a Contingent Valuation Method: The Case of the Korean DMZ." *Ecological Economics* 63, nos. 2–3: 511–520.

Lindsey, Peter, Robert Alexander, Johan Du Toit, and M.G.L. Mills. 2005. "The Potential Contribution of Ecotourism to African Wild Dog Lycaon Pictus Conservation in South Africa." *Biological Conservation* 123, no. 3: 339–348.

Lockie, Steward, Maree Franettovich, Vanessa Petkova-Timmer, John Rolfe, and Galina Ivanova. 2009. "Coal Mining and the Resource Community Cycle: A Longitudinal Assessment of the Social Impacts of the Coppabella Coal Mine." *Environmental Impact Assessment Review* 29, no. 5: 330–339.

Luning, Sabine. 2010. "Crisis What Crisis? Seeing Africa Like a Mining Company." Paper presented at the 2010 European Association of Social Anthropologists Conference, University of Maynooth, Ireland.

Mantz, Jeffrey. 2008. "Improvisational Economies: Coltan Production in the Eastern Congo." *Social Anthropology* 16, no. 2: 34–50.

McDonald, Kenneth. 2010. "The Devil is in the (Bio)diversity: Private Sector 'Engagement' and the Restructuring of Biodiversity Conservation." Antipode 42, no. 3: 513–550.

Moreno, Peter. 2005. "Ecotourism along the Meso-American Caribbean Reef: The Impacts of Foreign Investment." *Human Ecology* 33, no. 2: 217–244.

Myles, Peter. 2003. "Contribution of Wilderness to Survival of the Adventure Travel and Ecotourism Markets." USDA Forest Service Proceedings RMRS-P-27.

Nash, June. 1979. *We Eat the Mines and the Mines Eat Us: Dependency and Exploitation in Bolivian Tin Mines.* New York: Columbia University Press.

Novelli, Marina, Jonathan Barnes, and Michael Humavindu. 2006. "The Other Side of the Ecotourism Coin: Consumptive Tourism in Southern Africa." *Journal of Ecotourism* 5, nos. 1–2: 62–79.

Okazaki, Etsuko. 2008. "A Community-Based Tourism Model: Its Conception and Use." *Journal of Sustainable Tourism* 16, no. 5: 511–529.

Orlove, Benjamin, and Steve Caton. 2010. "Water Sustainability: Anthropological Approaches and Prospects." *Annual Review of Anthropology* 39: 401–415.

Pegg, Scott. 2005. "Can Policy Intervention Beat the Resource Curse? Evidence from the Chad-Cameroon Pipeline Project." *African Affairs* 105, no. 418: 1–25.

Powell, Robert, and Sam Ham. 2008. "Can Ecotourism Interpretation Really Lead to Pro-Conservation Knowledge, Attitudes, and Behavior? Evidence from the Galapagos Islands." *Journal of Sustainable Tourism* 16, no. 4: 467–489.

Prins, Harold. 1997. "The Paradox of Primitivism: Native Rights and the Problem of Imagery in Cultural Survival Films." *Visual Anthropology* 9, nos. 3–4: 243–266.

Reyna, Stephen, Andrea Behrends, and Günther Schlee (eds.). 2011. *Crude Dominations: An Anthropology of Oil.* Oxford: Berghahn Books.

Reyna, Stephen. 2011. "Constituting Domination/Constructing Monsters: Imperialism, Cultural Desire and Anti-Beowulfs in the Chadian Petro-State." Pp. 12–65 in *Crude Domination: An Anthropology of Oil,* ed. S. Reyna, A. Behrends, and G. Schlee. Oxford: Berghahn Books.

Robins, Steven. 2003. "Response to Adam Kuper's 'The Return of the Native.'" *Current Anthropology* 44, no. 3: 398–399.

Ross, Michael. 1999. "The Political Economy of Resource Curse." *World Politics* 51, no. 2: 297–322.

Ruiz-Ballesteros, Esteban. 2011. "Social-Ecological Resilience and Community-Based Tourism: An Approach from Agua Blanca, Ecuador." *Tourism Management* 32, no. 2: 655–666.

Ryan, Chris, and Jan Saward. 2004. "The Zoo as Ecotourism Attraction—Visitor Reactions, Perceptions and Management Implications: The Case of Hamilton Zoo, New Zealand." *Journal of Sustainable Tourism* 12, no. 3: 245–266.

Sachs, Jeffrey D., and Warner, Andrew. 1995. "Natural Resource Abundance and Economic Growth." National Bureau of Economic Research Working Paper #5398. Cambridge, MA.

Sawyer, Suzana. 2004. *Crude Chronicles: Indigenous Politics, Multinational Oil, and Neoliberalism in Ecuador.* Durham: Duke University Press.

Scheyvens, Regina. 1999. "Ecotourism and the Empowerment of Local Communities." *Tourism Management* 20, no. 2: 245–249.

Scott, James. 1998. *Seeing Like a State: How Certain Schemes to Improve the Human Condition Have Failed.* New Haven: Yale University Press.

Shandro, Janis, Marcello Veiga, Jean Shoveller, Malcolm Scoble, and Mieke Koehoorn. 2011. "Perspectives on Community Health Issues and the Mining Boom-Bust Cycle." Resources Policy 36, no. 2: 178–186.

Silver, Ira. 1993. "Marketing Authenticity in Third World Countries." *Annals of Tourism Research* 20, no. 2: 302–318.

Slater, Candace. 2002. *Entangled Edends: Visions of the Amazon.* Berkeley: University of California Press.

Smith, Timothy. 2012. "Crude Desires and the Pleasures of Going Green: Indigenous Development and oil extraction in Amazonian Ecuador." Paper presented at the European Association for Social Anthropologists Biannual Conference, Nanterre, France.

Stem, Caroline, James Lassoie, David Lee, and David Deshler. 2003. "How 'Eco' Is Ecotourism? A Comparative Case Study of Ecotourism in Costa Rica." *Journal of Sustainable Tourism* 11, no. 4: 322–347.

Stephan, Petra. 2000. "Sustainable Use of Biodiversity—What We Can Learn from Ecotourism in Developing Countries." Pp. 44–60 in *International Workshop: Case Studies on Sustainable Tourism and Biological Diversity,* ed. L. Gundling, H. Korn, and R. Specht. Bonn, Germany: German Federal Agency for Nature Conservation.

Stonich, Susan. 2006. "Enhancing Community Based Development and Conservation in the Western Caribbean." Pp. 77–86 in *Anthropological Contributions to Travel and Tourism: Linking Theory with Practice,* ed. T. Wallace. Washington DC: American Anthropological Association.

Stronza, Amanda. 2001. "Anthropology of Tourism: Forging New Ground for Ecotourism and Other Alternatives." Annual Review of Anthropology 30: 261–283.

Stronza, Amanda. 2005. "Hosts and Hosts: The Anthropology of Community-Based Ecotourism in the Peruvian Amazon." *National Association for Practice of Anthropology* 23: 170–190.

Stronza, Amanda. 2010. "Commons Management and Ecotourism: Ethnographic Evidence from the Amazon." *International Journal of the Commons* 4, no. 1: 56–77.

Stronza, Amanda, and Javier Gordillo. 2008. "Through a New Mirror: Reflections on Tourism and Identity in the Amazon." *Human Organization* 67, no. 3: 244–257.

Sullivan, Sian. 2006. "The Elephant in the Room? Problematizing 'New' (Neoliberal) Biodiversity Conservation." *Forum for Development Studies* 33, no. 1: 105–135.

Taussig, Michael. 1980. *The Devil and Commodity Fetishism in South America.* Chapel Hill: University of North Carolina Press.

Taussig, Michael. 1987. *Shamanism, Colonialism, and the Wild Man: A Study in Terror and Healing.* Chicago: University of Chicago Press.

Thomlinson, Eugene, and Donald Getz. 1996. "The Question of Scale in Ecotourism : Case Study of Two Small Ecotour Operators in the Mundo Maya Region of Central America." *Journal of Sustainable Tourism* 4, no. 4: 183–200.

Tobias, Dave, and Robert Mendelsohn. 1991. "Valuing Ecotourism in a Tropical Rainforest Reserve." *Ambio* 20, no. 2: 91–93.

Tsing, Anna Lowenhaupt. 2003. "Natural Resources and Capitalist Frontiers." *Economic and Political Weekly* 38, no. 48: 5100–5106.

Tsing, Anna Lowenhaupt. 2004. *Friction: An Ethnography of Global Connection.* Princeton: Princeton University Press.

Urkidi, Leire. 2011. "The Defence of Community in the Anti-Mining Movement of Guatemala." *Journal of Agrarian Change* 11, no. 4: 556–580.

Vivanco, Luis. 2006. *Green Encounters: Shaping and Contesting Environmentalism in Rural Costa Rica.* New York: Berghahn Books.

Wadley, Reed, and Michael Eilenburg. 2005. "Autonomy, Identity and 'Illegal' Logging in the Borderland of West Kalimantan, Indonesia." *Asia Pacific Journal of Anthropology* 6, nos. 2–3: 19–34.

Walsh, Andrew. 2003. "'Hot Money' and Daring Consumption in a Northern Malagasy Mining Town." *American Ethnologist* 30, no. 2: 290–305.

Walsh, Andrew. 2010. "The Commodification of Fetishes: Telling the Difference Between Natural and Synthetic Sapphires." *American Ethnologist* 37, no. 1: 98–114.

Walsh, Andrew. 2012. "Made in Madagascar: Speculating in and about Natural Wonders on 'The Real Treasure Island.'" Lecture delivered at Leiden University, 30 January.

Watts, Michael. 1999. "Petro-Violence: Some Thoughts on Community, Extraction, and Political Ecology." Institute of International Studies, of California–Berkeley, Working Paper WP99-1-Watts, 24 September 24.

Watts, Michael. 2004. "Resource Curse? Governmentality, Oil and Power in the Niger Delta, Nigeria." *Geopolitics* 9, no. 1: 50–80.

Watts, Michael. 2008. "Blood Oil: The Anatomy of a Petro-Insurgency in the Niger Delta." *Focaal* 52: 18–38.

Weaver, David. 2005. "Comprehensive and Minimalist Dimensions of Ecotourism." *Annals of Tourism Research* 32, no. 2: 439–455.

Weaver, David, and Laura Lawton. 2007. "Twenty Years On: The State of Contemporary Ecotourism Research." *Tourism Maganement* 28, no. 5: 1168–1179.

Welker, Marina. 2009. "'Corporate Security Begins in the Community': Mining, the Corporate Social Responsibility Industry, and Environmental Advocacy in Indonesia." *Cultural Anthropology* 24, no. 1: 142–179.

Werthmann, Katja. 2003. "Cowries, Gold and 'Bitter Money': Gold-Mining and Notions of Ill-Gotten Wealth in Burkina Faso." *Paideuma* 49: 105–124.

Weszkalnys, Gisa. 2009. "The Curse of Oil in the Gulf of Guinea: A View from Sao Tome and Principe." *African Affairs* 108, no. 433: 679–689.

West, Paige. 2006a. *Conservation Is Our Government Now: The Politics of Ecology in Papua New Guinea.* Durham: Duke University Press.

West, Paige. 2006b. "Environmental Conservation and Mining: Between Experience and Expectation in the Eastern Highlands of Papua New Guinea." *Contemporary Pacific* 18, no. 2: 295–313.

West, Paige, and James Carrier. 2004. "Ecotourism and Authenticity: Getting Away From It All?" *Current Anthropology* 45, no. 4: 483–498.

Wight, Pamela. 1996. "North American Ecotourists: Market Profile and Trip Characteristics." *Journal of Travel Research* 34, no. 4: 2–10.

Wight, Pamela. 2001. "Ecotourists: Not a Homogeneous Market Segment." Pp. 37–62 in *Encyclopedia of Ecotourism,* ed. D. Weaver. Wallingford, UK: CAB International.

Wheeler, Brian. 1991. "Tourism's Troubled Times: Responsible Tourism Is Not the Answer." *Tourism Management* 12, no. 2: 91–96.

Whitten, Norman, and Dorothea Whitten. 2007. *Puyo Runa: Imagery and Power in Modern Amazonia.* Urbana: University of Illinois Press.

Wunder, Sven. 2000. "Ecotourism and Economic Incentives: An Empirical Approach." *Ecological Economics* 32, no. 3: 465–479.

Zalik, Anna. 2010. "Oil 'Futures': Shell's Scenarios and the Social Constitution of the Global Oil Market." *Geoforum* 41, no. 4: 553–564.

Zimmerman, Erich. 1933. *World Resources and Industries: A Functional Appraisal of the Availability of Agricultural and Industrial Resources.* New York: Harper & Brothers.

Zografos, Christos, and David Allcroft. 2007. "The Environmental Values of Potential Ecotourists: A Segmentation Study." *Journal of Sustainable Tourism* 15, no. 1: 44–66.

Unintended Consequences

Climate Change Policy in a Globalizing World

Yda Schreuder

■ **ABSTRACT:** The cap-and-trade system introduced by the European Union (EU) in order to comply with carbon emissions reduction targets under the United Nations Framework Convention on Climate Change Kyoto Protocol (1997) has in some instances led to the opposite outcome of the one intended. In fact, the ambitious energy and climate change policy adopted by the EU—known as the Emissions Trading Scheme (ETS)—has led to carbon leakage and in some instances to relocation or a shift in production of energy-intensive manufacturing to parts of the world where carbon reduction commitments are not in effect. EU business organizations state that corporate strategies are now directed toward expanding production overseas and reducing manufacturing capacity in the Union due to its carbon constraints. As the EU has been "going-it-alone" with mixed success in terms of complying with the Kyoto Protocol's binding emissions reduction targets, the net outcome of the ETS market-based climate change policy is more rather than less global CO_2 emissions.

■ **KEYWORDS:** cap-and-trade, carbon constraints, carbon leakage, energy-intensive industries, EU Emissions Trading Scheme

International energy and emissions reporting entities present a bleak picture of the future of climate change and project that a pathway to long-term stabilization of greenhouse gas (GHG) emissions concentrations in the atmosphere at around 450 parts per million will likely not be achieved. In the 2008 *World Energy Outlook*—published annually by the International Energy Agency (IEA)—global energy demand is projected to grow by 1.6 percent per year on average until 2030, which translates into an increase of 45 percent over the same period. China and India are expected to account for over half of the increase in energy demand while the Middle East will emerge as a major new demand region (IEA 2008). The scenarios of future energy use as presented by the Intergovernmental Panel on Climate Change (IPCC) in 2007 provide some hope for mitigating emissions of GHGs but the apparent failure of the United Nations Framework Convention on Climate Change (UNFCCC) under the Kyoto Protocol in curbing GHG emissions creates doubt about how effective emissions reduction policies are (IPCC 2007).

Of the greenhouse house gases, CO_2 is the major gas responsible for global warming because of the immense volume in which it is emitted mainly from burning fossil fuel.[1] Although some of the trends predicted prior to the onset of the 2008 global financial crisis may have changed somewhat, the fact that most of the growth of fossil fuel use will occur in non-Organization for Economic Cooperation and Development (OECD) countries suggests that the overall predictions of climate change—a 6-degree Celsius rise by 2050 according to the 2012 IEA report—will

Environment and Society: Advances in Research 3 (2012): 103–122 © Berghahn Books
doi:10.3167/ares.2012.030107

not significantly alter in the next few decades (BP Energy Outlook 2011). In fact, under current policies, a doubling of CO_2 emissions worldwide by 2050 is not unlikely. The Global Carbon Project (2010) reports that carbon dioxide emissions rose 5.9 percent in 2010. The increase—a half billion extra tons of carbon emitted into the atmosphere—stands as the largest increase in any year since the Industrial Revolution and the largest percent increase since 2003. Researchers suggest that the high growth rate reflects a bounce-back from the 1.4 percent drop in emissions in 2009—the year the recession had its biggest impact—but the Global Carbon Report suggests that little progress has been made in limiting greenhouse gas emissions reduction and that combustion of coal accounts for more than half of the growth in emissions. In China alone, emissions grew 10.4 percent in 2010, mostly attributed to the use of coal.

Debated presently—but not seriously considered at the time when the Kyoto Protocol was first signed in 1997—is the recognition that national emissions reduction commitments have little effect in a global economy driven by a rapid increase in foreign trade and foreign investment and organized around multinational corporations and international production networks (Giddens 2009; Klein 2008; Koch 2011; Newell and Paterson 2011; Schreuder 2009). This naturally leads to questions about the merits of a market-based regulatory approach to curbing CO_2 emissions as adopted by the Kyoto Protocol and implemented by the EU in 2005. As the Kyoto Protocol was based on the premise that developed countries would assume the largest responsibility for the concentration of accumulated GHG emissions in the atmosphere, we now know that a far greater share of CO_2 currently emitted derives from production in emerging economies; India and China in particular.

This article assesses the forces that drive the global market economy and determine why multinational corporations and in particular energy-intensive industries strategize to relocate or outsource production to developing countries that are not subjected to GHG emissions reduction targets under the Kyoto Protocol. The discussion will focus on carbon leakage and the expanded use of coal as a source of energy for electricity generation and manufacturing in developing countries. The expanding economies of India and China, in particular, warrant specific attention because of their rapid growth in industrial capacity. The article's title "Unintended Consequences" refers to the notion that contrary to the Kyoto Protocol's objective, global GHG emissions have increased rather than been reduced, and, ironically, the implementation of the EU Emissions Trading Scheme (ETS) or cap-and-trade system may be partially to blame. Carbon leakage is the term used to describe how, due to emissions reduction policies in one part of the world, enhanced emissions occur in other parts of the world with the net effect that overall global emissions increase. The assessment of the impact of the implementation or the ETS on carbon leakage is presented in the context of the rapidly expanding global economy under the capitalist system of production since the 1980s when the World Trade Organization (WTO), the World Bank, and the International Monetary Fund (IMF) began to develop their policies under the guidance of neoliberal policies promoted in Washington and London. The article forms a case study of how market forces drive growth in the global economy and confront the limits of growth to the Earth resource base and carrying capacity.

Climate Change Policy in a Globalizing World

While emissions have stabilized in some of the developed economies, international trade and foreign direct investment by multinational corporations has shifted the source of CO_2 emissions to developing countries (Schreuder 2009). We now recognize that the rise in emissions from goods produced in developing countries (non-OECD) but consumed in industrialized countries (OECD) was six times greater than the emissions savings of industrialized countries (Peters et al. 2011) and that relocation or shift in production of particularly energy-intensive manufacturing played a key role in the redistribution of emissions.

The Kyoto Protocol recognized two groups of countries; Annex 1, which comprised most of the industrialized countries, and non-Annex I developing countries.[2] Annex I countries committed to legally binding reductions of greenhouse gas emissions of—on average—5.2 percent below 1990 levels between 2008 and 2012, with the US target set at 7 percent and the EU target set at 8 percent. The Kyoto Protocol came into force on 16 February 2005 after 55 Annex I countries covering at least 55 percent of 1990 GHG emissions had ratified the treaty.[3] The United States did not ratify the Kyoto Protocol and the Bush Administration officially withdrew from the Kyoto Protocol shortly after the president's inauguration in 2002. Russia was the last major country to ratify the Protocol in November 2004. India, China, and Brazil ratified the treaty but were not required to commit to emissions reduction targets under the Kyoto Protocol as they were classified as non-Annex I countries.

Whether it was fair or expedient to not submit developing countries to binding GHG emissions reduction efforts under the Kyoto Protocol has been discussed since the beginning of the UNFCCC in Rio de Janeiro in 1992. Clearly, as the largest share of historical greenhouse gases originated in developed industrialized countries and as the per capita emissions in developing countries were still relatively low, developing countries were not eager to commit to binding emissions reduction targets during the Kyoto Protocol's commitment period (2008–2012) but should—according to the main parties of the UNFCCC—be encouraged to participate in a meaningful and eventually significant way to help curb global emissions. From the start of the negotiations in Rio de Janeiro, it was understood that the share of global emissions originating in developing countries would grow as these countries implemented economic development policies to meet their social and development needs. Annex I countries agreed that—as developing countries develop industrial capacity—they would help pay for and supply technologies to them for climate change–related projects in order to encourage a more energy-efficient and lower-emissions development path. This clause in the Kyoto Protocol was defined as the Clean Development Mechanism (CDM) and through the linking of the EU Emissions Trading Scheme (ETS) with the Kyoto Protocol, the CDM has been applied by the ETS as one of the major mechanisms by which countries could meet their agreed-to emissions reduction commitments.[4]

Critics of the Kyoto Protocol—including US government officials under the Bush Administration—had always maintained that since China, India, and other developing countries would soon be among the countries contributing a major share to global GHG emissions, they should come on board with binding GHG emissions reduction commitments for the post-Kyoto negotiations. Without carbon constraints imposed on developing countries, they argued, corporate industries in developed countries would lose their competitive position and would likely expand production in non-carbon-constrained countries like China, India, and Brazil, or any other country competitively positioned for foreign direct investment (FDI) or foreign trade. In that case, there would be no net reduction of GHG emissions concentrations in the atmosphere but just a shift in the geographical distribution of the source of emissions due to expanding manufacturing capacity in the developing non-Annex I countries. The net result would be carbon leakage, which suggests an increase in overall global carbon emissions as a result of emissions reduction strategies and legislation in countries where climate change policies aimed at CO_2 emissions reductions apply.[5] Whereas this may sound contradictory, in fact it is very logical if we consider competitive forces in the global economy. Because CO_2 reduction policies will likely increase production costs in countries where abatement strategies are in effect, the market would shift production to nonabatement countries. This would be particularly the case for energy-intensive industries, which are defined as industries that use a relatively large amount of energy per unit value manufactured and therefore produce high emissions relative to its useful output.

The market-based approach to climate change policy as designed under the Kyoto Protocol and as later implemented by the EU was heavily influenced by political considerations at a time when neoliberal policies—and in particular US policies—dominated international debates and negotiations. The World Commission on Environment and Development (WCED 1987) had issued the so-called Brundtland Report, *Our Common Future,* which promoted science and technology and a free-market approach to sustainable development. In preparation for the UN Conference on Environment and Development (UNCED) also known as the Earth Summit, in Rio de Janeiro in 1992, the Business Council for Sustainable Development —an international group of CEOs representing the major global corporations—was formed to advise UNCED on business and industry issues and to stimulate involvement by business in UNCED. Its leader, Maurice Strong, became the secretary-general to chair the UNCED negotiations. Environmentalists saw the Business Council as representative of business interests in policy making and as evidence of corporate hijacking of UNCED. Within a year of the Earth Summit, The Ecologist's edited volume, *Whose Common Future?* (1993) questioned the success and credibility of the negotiations. Even though the meetings were perceived to be all-inclusive and broad-ranging, it was clear that the corporate sector was a major player both in the formulation of the various conventions and as actor in the negotiations. In the battle to save the planet, free market environmentalism was promoted and the corporate sponsors were given special access to the secretariat. The philosophy of the Business Council on Sustainable Development prevailed throughout most of the deliberations and the desirability of economic growth, the market economy, and the Western development model based on neoliberal principles were not questioned. UNCED thus never had a chance of addressing the real problems of the environment and development relationship, according to the authors of *Whose Common Future?* (1993). The Earth Summit's action plan, *Agenda 21,* suggested ways to enable poor nations to achieve sustainable development, but did not question the desirability of the rich nations' pursuit of the same. So, the authors of *Whose Common Future?* (1993) asked the question in whose interest we are promoting sustainable development, and who is managing it?

The recommendations for sustainable development as adopted by the UNCED in 1992 followed fairly closely the recommendations made by the WCED. In the Brundtland Report commissioned by the WCED sustainable development is defined as development that meets the needs of the present generation without compromising the ability of future generations to meet their needs. In *Planet Dialectics* (1999), Wolfgang Sachs questioned the recommendations of the Brundtland Report and the support for the Western model of development as it is at odds with both equity considerations in the present generation as well as sustainability of the Earth resource base and carrying capacity considering future generations. Sachs poses that in a fundamental way sustainability is about global citizenship and argues that the principles of equity and sustainability derive from equal access to resources and the global commons. In his critique of the Earth Summit, Sachs claims that the call from the developing countries for a more equitable share in global wealth creation and access to resources was translated in terms of the right to development.

At the Earth Summit the leaders from the developing countries aligned with the business community from the developed countries in their praise for economic development as the solution to all global environmental problems. The argument was that with higher levels of economic development and technological know-how would come greater care for the environment and more efficient use of energy, thus lower pollution and GHG emissions levels. As Sachs (2002) later stated, the quest for justice was firmly wedded to the idea of development; nobody had to profoundly change, and all parties could turn to business-as-usual, a position amply borne out in recent years. The authors of *Whose Common Future?* (1993) asserted that the UN-sponsored Earth Summit was nothing more than a repeat of the development debates of the 1960s and 1970s. They maintained that mainstream solutions proposed at the Earth Summit would be counterproductive because the Western economic development model was never questioned. The US and EU proposals presented at the Earth Summit to combat global environmental crises, including the UNFCCC climate change policy proposal, recommended limiting population growth, stimulating free-market enterprise, and the application of Western technology and transfer of this technology, know-how, and capital. The recommendations, according to the authors of *Whose Common Future?* (1993) did not sound terribly convincing after decades-long efforts to fight poverty, famine, and starvation by the same means. Repeated efforts to make life for the majority of the population of the developing world more tolerable through programs such as basic needs fulfillment, human resources development and education, and now sustainable development, obscure the real issue, which is that the introduction of the Western development model has more often than not resulted in increased poverty and environmental degradation (Harvey 2006; Peet 2007; Slater and Taylor 1999).[6]

Neoliberal Capitalism and the Limits to Growth

As Daly (1996) pointed out, economic growth can only proceed to the point where throughput of matter and energy stays within the regenerative or assimilative capacity of the Earth ecosystems. Thus, ultimately, sustainable development has to be understood as "development without growth" at a steady state, which requires a major geopolitical change and political-economic adaptation. In pursuit of profit from business-as-usual economic growth, many believe that capitalism in its present neoliberal form is unsustainable (Daly 1996; Harvey 2005, 2006; Klein 2008; O'Connor 1994; Porritt 2005; Schreuder 2009; Wallerstein 2000). Neoliberal capitalism aims to render maximum profit at minimum costs and thus global competition drives many businesses to outsource to locations where labor costs are lower and where strict environmental rules and regulation are not in effect (Chomsky 1999; Sen 1999; Stiglitz 2002). The absence of carbon constraints and the abundance of coal as cheap energy source form two other reasons why relocation may occur; a scenario causing much higher levels of CO_2 emissions than would have been the case had the Kyoto Protocol and the EU Emissions Trading Scheme not been in effect.

In essence, it can be argued that an economic system built on profit and global competition has caused an irreparable rift with the natural laws of life (Chomsky 1999; Harvey 2006; Klein 2008). Stern admits in *The Economics of Climate Change* (2006) that climate change policy under the UNFCCC Kyoto Protocol shows clear evidence of market failure. In his persuasive expose, *Capitalism as If the World Matters*, Porritt (2005) charges that the forces of capitalism and the challenges of the biophysical limits to growth require profound transformations if we want to avoid dramatic disruptions to life on earth. Arguing that capitalism in its present form is unsustainable, he asserts that only principles of sustainable development can provide the foundations upon which to base the transformations necessary to the global challenges we now face. He suggests that core values like a sense of interdependence, empathy, equity, personal responsibility, and intergenerational justice should be the guiding principles for a new world vision, but how to channel national interests, individualism, materialism, greed, and pursuit of riches, into a more sustainable lifestyle, remains the question. Such a transformation requires a major adjustment or change from our business-as-usual approach to development and economic growth.

The US directed neoliberal model of development has been the dominant model of global economic development since World War II (Sachs 1999). Western development strategies and capitalist interests and pursuits have become the guiding principles of most international negotiations (including the UNFCCC) and UN institutions like the WTO, the World Bank, and the IMF. These institutions, initiated and supported by the US and the EU have formed the mechanism by which foreign investment and trade have been promoted. The so-called neoliberal principles have directed development—in particular since the days of Ronald Reagan and Margaret Thatcher—under the banner of the Washington Consensus. The term Washington Consensus was first coined in the late 1980s by John Williamson (1990) of the Institute for International Economics; a research institution devoted to the study of international economic policy to promote a relatively specific set of economic policy prescriptions that were considered to constitute a standard reform package recommended for countries in economic distress by Washington-based institutions such as the IMF, the World Bank, and the US Treasury Department. The policies the Washington Consensus promoted were policies pursued in Chile after the fall of President Salvador Allende (Valdez 1995) and the oil crises of the 1970s but they became a set of quite specific prescriptive dictates during the 1980s with the debt crisis in Latin America when the IMF and World Bank signed on to them (Harvey 2006; Sachs 1999).[7] The term neoliberalism refers to an intellectual and political movement that espouses economic liberalism as a means to promoting economic development and securing political liberty. Inspired by Friedrich Hayek and Milton Friedman, the movement is sometimes described as an effort to revert to the economic policies of nineteenth-century classical liberalism based on Adam Smith's and David Ricardo's ideas of national economic growth but, more specifically, it refers to the historically specific reemergence of economic liberalism among economists and policymakers during the 1970s through 1990s—the period of the Washington Consensus.[8]

Neoliberalism is not a unified economic theory or political philosophy but it denotes neoclassical influenced economic approaches and libertarian political philosophies that portray government control over the economy as inefficient, corrupt, or otherwise undesirable. It largely rejects post–World War II Keynesian economics and—as pursued by the WTO, the World Bank, and the IMF—has greatly advanced the interests of multinational corporations. Also, neoliberal economic policies have led to rapid expansion of foreign investment in and trade with emerging markets like China and India. As a result, both India and China have entered a period of rapid economic growth. In the case of India, outsourcing of information technologies (IT) and financial services played an important role in the growth spurt. In China, rapid growth occurred in industrial production, much of it through investments by multinational corporations for export production. China's real gross domestic product (GDP) has grown at around 10 percent per year for the past decade and economic forecasts for further growth remain strong. Together with strong economic growth, consumer demand is surging in both India and China because of the growing affluence of an emerging urban middle class. With China's entry into the WTO in November 2001, the Chinese government made many specific commitments to trade and investment liberalization that have substantially opened up the Chinese economy to investment by foreign firms through FDI and foreign trade. Much of the investment has occurred in the energy-intensive sector, which has contributed to the rapid increase in carbon emissions (EIA 2006).[9]

In both India and China, manufacturing capacity is fueled mostly by direct combustion of coal. Predictions are that the growing energy demand could drive a fourfold increase in the use of coal by 2030, which would result in a greatly increased annual emissions total. Together with strong economic growth and increased consumer demand, India and China's demand for fluid fuels is surging as well. The EIA (2006) forecasts that China's oil consumption will increase by almost half a million barrels per day, or over 40 percent of the total growth in world oil demand in the next decade.[10] Other rapidly growing developing countries in Latin America, such as Brazil and Argentina, or Mexico and Chile, are also fast becoming large energy users and carbon emitters. In terms of the increase in pressure on the carrying capacity of the Earth ecological system as a result of the rapidly growing global economy, important questions arise as to how the world's ecosystems can withstand the ongoing increase in carbon emissions in the atmosphere. The IPCC (2007) alerts us to the fact that carbon dioxide concentration in the atmosphere has increased from a preindustrial value of about 280 parts per million to a rapidly approaching critical figure of 450 parts per million. The 2005 Millennium Ecosystem Assessment conducted under the auspices of the United Nations concluded that the world's ecosystems, ranging from water, soils, the biosphere, and the atmosphere, are seriously undermined as the world's rapidly developing countries are moving to center stage of the global economy (World Health Organization 2005).

The EU Emissions Trading Scheme and Carbon Leakage

The best example of a failure of a market-based approach to climate change policy is the EU ETS implementation of cap-and-trade. The implementation of the EU ETS provides an opportunity to assess how the forces of corporate capitalism address and deal with carbon emissions reduction in a partially carbon constrained global economy. In analyzing the impact of EU climate change policy, I focus on carbon leakage and the more prominent place of coal in fuelling manufacturing industries around the world leading to increased global emissions.

The EU ETS was implemented in 2005 and forms the cornerstone of the EU climate change policy covering about 45 percent of total EU CO_2 emissions. The ETS became the showcase of the EU commitment to reduce GHG emissions. Since the EU ETS is linked to the Kyoto Protocol's Flexible Mechanisms, it also has a global reach and is considered the model for climate change policy worldwide.[11] The ETS has been studied and analyzed by business groups and environmental groups alike as both groups tried to influence policy makers by showing how effective or ineffective or damaging the ETS was going to be on global emissions and industrial competitiveness. Many considered the first phase of the EU ETS (2005–2007) as an experiment or test case but as we are now approaching the end of the second phase of the ETS (2008–2012), it is time to make up the balance sheet.[12]

Emissions trading or cap-and-trade was chosen as it promised to meet the EU GHG emissions reduction goal in the most cost-effective way (Ellerman and Joskow 2008; Hillebrand et al. 2002). The ETS only covers CO_2 emissions from large industrial and energy installations in a limited number of energy-intensive sectors such as refining, coke ovens, cement, pulp and paper, glass, steel and metal, and power generation. By establishing a market price for carbon, EU policymakers envisioned that industrial firms in these sectors would make investments based on reducing emissions and improving energy efficiency. Combined with a robust compliance system, emissions trading would ensure that emissions reduction targets would be met and as such the ETS would comply with the implementation of the Kyoto Protocol targets. In balancing effectiveness and competitiveness, complex, and sometimes perverse incentives were entered into the scheme that contributed to a good deal of skepticism among the public.

The EU ETS introduced the national allocation of CO_2 allowances, which permit particular segments of industry to emit certain amounts of carbon dioxide. Each member state of the EU had to first submit a National Allocation Plan (NAP). Based on the agreed-to Kyoto reduction of CO_2 emissions, each EU member had established its own reduction target. In the NAP, member states could specify which industrial sectors would be covered and which would be excluded.[13] Furthermore, each country could specify how new entrants, closures and transfers or mergers would be treated, and what kind of allocation methodologies would be used. In the decision-making process, industrial representatives, environmental groups, and other interested stakeholders had a good deal of influence, and the ultimate reason why some sectors were included and others excluded was not always clear. Many believe that harmonization of the NAPs and better oversight is urgently needed for the EU ETS to gain credibility and for the scheme to become the model for CO_2 emissions trading. Even though the EU ETS will ultimately be judged based on its effectiveness as a tool to reduce GHG emissions, the underlying rationale for choosing emissions trading was based on economic considerations.

By implementing the EU ETS an attempt was made to account for a market externality (CO_2 emissions) with a minimal impact on competitiveness. In theory, the market price of carbon is driven by the abatement costs of CO_2 emissions reduction, ensuring that the target reduction is achieved at the least cost. By creating a market price for carbon, investment would be made in energy efficiency and better process technology. The ETS offered business flexibility to achieve the objective by low-cost abatement or by allowing credits from the Kyoto Protocol's Flexible Mechanisms to be used for compliance. Correct allocation of the number of permits per industry and installation was critical. The number of permits should be fewer than the CO_2 emissions reported in 1990, however, during the ETS first round most member countries allotted a far greater number of permits than would serve the carbon market. During phase I (2005–2007), allowances were allotted for free, which meant that there was hardly a price for carbon and therefore no incentive to invest in energy efficiency or improved process technology. At the end of the first phase of the ETS, the parties involved realized that the Kyoto Protocol CO_2 emissions reduction scheme was not on target, which led to an intensive debate on the effectiveness of issuing free allowances and the desirability of auctioning carbon credits during the second phase. By this time, the EU had begun a public debate on the wisdom of the EU going-it-alone policy as interested parties anticipated that higher energy costs would eventually affect the Union's global competitiveness and slow down investment in Europe with the subsequent expansion of production to non-Annex I countries or the United States. This response, known as carbon leakage, suggests that an increase rather than a reduction in global GHG emissions would be the result.

As discussed earlier, carbon leakage is the term used to suggest an increase in overall global carbon emissions as a result of emissions reduction strategies in countries where climate change policies aimed at CO_2 emissions reductions apply (Annex I) in instances where the producer cannot pass on the increase in cost of production to the consumer. This is the case, for instance, in aluminum production where prices are set on a global exchange, and where the manufacturer will want to locate production where carbon constraints are not in effect and where coal is abundant and cheap. Thus, CO_2 emissions reduction targets in some parts of the world may have an effect on demand for fossil fuels with higher carbon content in nonabatement countries (Sijms et al. 2004).

The lobbying group for energy-intensive industries began a public debate about the wisdom of the EU's going-it-alone policy from the very beginning, when the EU Commission began the deliberations about the ETS targeting energy-intensive industries in 2001 (Alliance of Energy Intensive Industries 2004, 2005). A cap-and-trade system to reduce carbon emissions would carry a carbon price and carbon would thus become a cost of production. Anticipating that the EU economy under the Kyoto regime would become a carbon constrained economy, industry leaders lobbied hard to give specific direction to the implementation of an emissions trading scheme. The Alliance of Energy Intensive Industries has been the main lobbying group for the industry. It was argued that goods that contained more carbon or had been produced with greater energy intensity would be relatively more expensive than goods that contained less carbon or used less energy. Therefore, in a partly carbon constrained global economy, carbon constrained countries would import goods from nonabatement countries where no carbon constraints applied. In November 2005, the Alliance issued a call for action on the part of the EU to resolve these fundamental problems associated with the rise in energy prices as the position of EU energy-intensive industries in the global market was seriously undermined (Alliance of Energy Intensive Industries 2005). As the EU manufacturing industry was paying the price for a hastily designed ETS, they argued, policymakers should take responsibility for the failure of the scheme and solve the problem by reforming the ETS. The Alliance recommended that CO_2 prices should be separated from power prices and that windfall profits on the part of power producers should not be at the expense of energy-intensive manufacturers.[14]

Producers of energy-intensive products such as iron and steel, and aluminum manufacturers have four choices in dealing with cost increases due to carbon constraints. First, they can invest in more energy-efficient plants or process technology. If this is not an option then they can buy allowances provided these are available on the carbon market at a reasonable price. If they cannot afford to buy allowances or carbon permits to facilitate production then future business prospects will be affected and market share of the company will fall. The fourth and final option is to relocate production outside the carbon-constrained region. The latter, obviously, is the most damaging as carbon leakage occurs mainly between Annex I and non-Annex I countries of the UNFCCC and between those Annex I countries that have committed to CO_2 emissions reductions under the Kyoto Protocol and Annex I countries that did not ratify the Protocol and therefore did not commit to binding emissions reduction targets (e.g., the United States). Carbon leakage may also occur among committed Annex I countries with high reduction targets (like the West European EU members) and countries such as Russia, the Ukraine, and some East-Central European countries, which experienced a decline in emissions due to stagnating economic performance after the fall of communism (see Table 1).

Carbon leakage can be triggered by direct carbon costs (price for carbon allowance or carbon credits) and indirect carbon costs resulting from higher power or electricity prices. Carbon leakage is likely to occur if carbon costs are high and cannot be passed on to the consumer via higher product prices and if production is exposed to international competition and foreign

Table 1. Comparison of Greenhouse Gas Emissions, 2005 (total, per capita, per GDP)

	Total (in Mt CO_2 equiv.*)	Per Capita (in tons CO_2 equiv.)	Per GDP (in tons CO_2 equiv. per thousand US$)	Percent Change in total greenhouse gas emissions (1990–2005; %)
China**	4,963.1	3.9	4.1	31.7
European Union	4,953.5	10.7	0.5	−5.8
Russian Federation	2,289.2	15.9	6.5	−23.0
United States	7,241.5	24.2	0.7	18.6

* Mt CO2 Equiv.: Million metric tons of CO2 equivalent.
** China figures based on data from 2000.

Sources: Population: UN, Eurostat; GDP: IMF, World Bank; Greenhouse Gas Emissions: UNFCCC, WRI. Data compiled by: Econsense (2007).

trade (Alliance of Energy Intensive Industries 2004, 2005). Carbon leakage is less likely to occur if the costs of carbon credits can be passed on to consumers or if products are highly specialized and serve a regional market.

The occurrence of carbon leakage under the Kyoto Protocol is usually expressed as a percentage of CO_2 emissions increase that results from an increase in emissions in a nonabating country divided by the reduction of emissions by a country subject to an emissions reduction target under the Kyoto Protocol. Thus a 20-percent carbon leakage rate means that 20 percent of reductions in emissions in an abatement country are reversed as a result of increased emissions elsewhere. Measurement of carbon leakage is not an exact science as the increase in CO_2 emissions in any one country as the result of CO_2 abatement in another country is difficult to separate from other factors, like lower labor costs, that may determine a market shift or a shift in CO_2 emissions. Still, by understanding the mechanisms through which carbon leakage can occur, it is possible to assess the impact of climate change policies to some extent.

Various simulation models have been developed and numerous studies calculating CO_2 induced costs increases have been made by industries and affected parties (Sijms et al. 2004). Most estimates are derived from Computable General Equilibrium (CGE) models, which are economic models that use actual economic data to estimate how an economy might react to changes in policy, technology, or other external factors. The equations often assume cost-minimizing behavior by producers and consumers. A CGE model database consists of tables of transaction values, showing, for example, the value of coal used by the iron and steel industry and is usually presented as an input-output table. However, these are static models and cannot forecast the future. To account more fully for the effects of the anticipated market changes and geographical distribution of energy-intensive production, trade, and carbon leakage, models incorporating strategic interaction among firms producing energy-intensive products have been developed using hybrid datasets based on the Global Trade Analysis Project (GTAP) and energy balances, prices, and taxes derived from IEA accounts, which created so-called GTAP-E models (Babiker 2005; Weyant 1999).[15] Results from the different models and estimates vary greatly and have been the source of much controversy with regard to the economic impact of the implementation of the Kyoto Protocol (Stern 2006). The typical CGE model estimated values of carbon leakage due to the implementation of the Kyoto Protocol are at between 5 percent and 25 percent worldwide but some GTAP/IEA dataset-based models or GTAP-E models predict carbon leakage rates as high as 130 percent (Babiker 2005). In the latter case, the Kyoto Protocol would lead to a huge increase in global carbon dioxide emissions. In one study, it is estimated that the United States will be the largest contributor to carbon leakage in 2020 if the country decides not to participate in a Kyoto-type emissions trading scheme (Hamasaki 2007).

EU country and industry studies have identified the different industrial sectors most immediately affected by carbon abatement policies (Hamasaki 2007; Sato 2007; Weinreich 2009). These studies calculated and identified both direct and indirect CO_2 induced costs increases for various industrial sectors and found that by far the highest costs increases were incurred by the lime and the cement industries (Weinreich 2009). However, because there is little or no international competition and foreign trade involved in these sectors, the impact of cost increases due to carbon pricing is insignificant. Less affected by the impact of carbon price but more vulnerable to international competition is the iron and steel sector especially in coastal locations. The aluminum and aluminum products sector as well as basic chemical products, pulp, copper, glass, dyes, and pigments are fully exposed to international competition and have experienced relocation effects of high carbon costs (Weinreich 2009). From these and other studies conducted, the impact of trade liberalization is viewed as a major contributing factor to carbon leakage (Kuik and Gerlagh 2003; Sijms et al. 2004; Taylor and Copeland 2005). Abatement measures in Annex I countries—under conditions of liberalized trade—changes geographical production and consumption patterns of manufacturing and energy production and will in effect increase CO_2 emissions of non-Annex I and nonabatement countries through enhanced international trade and global investment flows (CRU News, March 2006).[16]

Coal as a Global Energy Source

The 2008 market report of EURACOAL—the industry association representing the European coal industry—documents that global coal production increased by over 200 million tonnes (mt) during 2008, most of which was mined in China. In contrast, EU production was down from previous years and Europe was the only region where coal production is decreasing according to EURACOAL (2009). In fact, over the past decade, EU production has fallen by 35 percent in the expanded EU-25 region and by 50 percent in the EU-15 region (see Table 2).[17] At the same time there has been an upsurge in EU coal import of 40 percent in just ten years. The combined effect of EU regulations governing state aid for the coal industry, which expired in 2010, and the EU ETS, which commits some 12,000 energy-intensive plants to buy carbon permits, are the major reasons for the EU decrease in coal production and use for electricity generation. Coal emits approximately twice as much CO_2 as natural gas and the-cap-and trade regime favors the use of gas in electricity generation as EU industrial emissions are capped, since 2008, at 20 percent below 2005 levels by 2020 in order to reduce emissions more quickly.[18]

At the same time, the United States, China, and Russia are all increasing coal production and are holding among the largest coal reserves, which suggests that if they do not commit to drastic and binding emissions reduction targets, sharp increases in coal production and coal use can be expected. In 2009, the prospect of regulating emissions via carbon trading—as the Obama Administration promoted—alarmed coal-producing and steel-producing states in the American Midwest, and congressmen from these states actively rallied against the Obama plan. Presently, supporters of reducing GHG emissions in the United States bank their hopes on regulation of emissions through the Environmental Protection Agency.[19] Meanwhile, for as long as Russia, China, India, and other emerging economies are on the sidelines with respect to climate change policy it is unlikely that the US federal government will go much further than what is currently being legislated. Furthermore, it is unlikely that the EU going-it-alone cap-and-trade policy will continue without a firm international Kyoto-type climate change policy in place as unilateral climate change policy greatly diminishes the EU's global market position.[20]

Table 2. Coal Production and Consumption at the end of 2008

	Production			Consumption		
	1998	Mt oil equiv. 2008	Change 2008 over 1998 (%)	1998	Mt oil equiv. 2008	Change 2008 over 1998 (%)
United States	603.2	596.9	−1.0	545.7	565.0	3.5
Russian Federation	103.9	152.8	47.0	100.7	101.3	−0.6
China	628.7	1414.5	125.0	651.9	1406.3	115.7
European Union	229.2	171.5	−25.2	323.5	301.2	−6.9

Source: Data compiled from British Petroleum, *Statistical Review of World Energy* (2008).

Both Russia and China have greatly expanded coal-generated electricity capacity. Today, most of Russia's energy is generated from gas but Russia's domestic demand for coal is increasing and the country's government has decided to reduce gas consumption in domestic power generation and increase coal use in order to maximize gas exports. Russia trails China, the United States, Indonesia, and Australia in terms of coal output but it is expected that the country will heavily invest in improving the electric grid infrastructure and rail and port facilities in order to improve the use of coal for energy generation. As a result, Russia's supply of coal for domestic use will likely rise from a current 130 mt per year to between 250 and 325 mt by 2020.[21] This will make Russia attractive for investment by foreign companies in energy-intensive industrial sectors and is likely going to contribute to more carbon leakage.

Coal currently fuels over 40 percent of electricity worldwide and will play a vital role in electricity generation in the next few decades according to information from the World Coal Institute (2009). With availability of abundant, affordable, and geographically dispersed reserves, coal is considered a secure and reliable source of energy worldwide. Coal prices have historically been lower and more stable than oil and gas prices and despite an increase in volatility of the energy market this has essentially remained the case. Coal is therefore likely to be the most affordable and reliable source of fossil fuel for power generation in many developing and industrialized countries in the foreseeable future. For energy-intensive industries, the impact of fuel and electricity price increases and price volatility will have important implications for location decisions and countries with access to indigenous energy supplies and affordable fuels from a well-supplied market can avoid volatility and uncertainty, enabling them further economic development and growth potential from a manufacturing industry.

International trade in coal reached 917 mt in 2007 accounting for about 17 percent of the total amount of coal consumed (World Coal Institute 2009). Australia is the world's largest coal exporter with over 244 mt of coal in 2007 exported out of its total production of 323 mt. Australia is also the largest supplier of coking coal —used in iron and steel production—accounting for 53 percent of world exports. Consumption or use of coal for electricity generation is projected to grow by 1.5 percent per year until at least 2030. Coking coal is more expensive than coal used in electricity generation, which means that Australia is able to afford the high freight rates involved in shipping coking coal over long distances. The largest market for Australian coal is Asia, which currently accounts for over 50 percent of global coal consumption (World Coal Institute 2009). Although China imports the largest share of Australian coal, other Asian countries that do not have carbon fuel resources sufficient to cover their energy needs, also import Australian supplies to help meet their demand. Japan, Taiwan, and Korea, for example, import significant quantities of coal for electricity generation and coking coal for steel production (World Coal Institute 2009). According to *BP's Statistical Review of World Energy* (2008), coal has been the fastest-growing major fuel and coal consumption grew by 3.1 percent in 2008. China's share of world energy consumption growth in 2007 was 52 percent much of it derived from coal with more than two-thirds of global growth in coal consumption attributed to increase in coal consumption in China (see Table 2).

China has entered a period of rapid economic growth based on industrial investments by multinational corporations for export production (Li 2008; Zang and Pearce 2012). Together with strong economic growth that increases energy demand, consumer demand is surging also because of the growing affluence of an emerging urban middle class. With China's entry into the WTO, the Chinese government made specific commitments to trade and investment liberalization, which substantially opened up the Chinese economy to investment by foreign firms. Much of the investment has occurred in the energy-intensive sector, which has contributed to the rapid increase in carbon emissions as manufacturing is fuelled mostly by direct combustion of coal. The EIA has predicted that China's growing energy demand could drive a fourfold increase in the use of coal by 2030, which would translate into the largest growth in global carbon dioxide emissions (EIA 2006: table A-10). As carbon-emitting industries multiply at a rapid rate, China is building on average one coal-fired power plant every week. Coal makes up approximately 70 percent of China's total primary energy consumption and the country was both the largest consumer and the largest producer of coal in the world in 2007 at 1,311 mt and 1,289 mt, respectively, or over 41 percent of the world total in both consumption and production of coal. In addition, China records 114.5 billion mt or 13.5 percent of the world total of proven coal reserves, which places it third in the world behind the United States and Russia (Table 2). Opening up to foreign investment and foreign trade gave China a competitive advantage over the more mature economies of the EU and the US. As the demand for energy has put China on the fossil-fuel coal-based development path, the investments made and the infrastructure developed will shape the manufacturing economy for several decades to come. China is currently not constrained by climate change policies and GHG emissions levels have increased over 30 percent above 1990 levels by 2005 (Table 1). It is likely that the trend will continue or resume after the global recession ends.

The Impact of Cap-and-Trade Policies on the Iron, Steel, and Aluminum Industries

If the metal industry trade publications are any clue, then the projected increases in coal consumption and carbon dioxide emissions are not far off. According to a report issued by the American Iron and Steel Institute in 2006, worldwide production of steel increased by about 470 million tonnes during the preceding decade, with most of the expansion occurring in countries that use less energy-efficient production methods and impose weaker environmental regulation or enforcement (i.e., fewer carbon constraints). China's share of world production of steel had almost tripled from 13 percent in 1996 to 35 percent in 2006, a 316 percent increase (American Iron and Steel Institute 2006). China's steel export also tripled, increasing 309 percent between 1995 and 2005, and China is now the largest steel-exporting country in the world. India, similarly, is rapidly expanding its steel production capacity and export. Therefore, the American Iron and Steel Institute report concludes that if measures to control carbon emissions do not take international trade and environmental costs into account, then total global carbon emissions will increase dramatically (see also Sato et al. 2007).[22] Meanwhile, the EU is the only major region to show a drop in export of steel. The European Confederation of Iron and Steel Industries (EUROFER 2009) reports that Europe's steel production was only 198 mt in 2009 while China's production was more than double that at more that 70 percent of the growth in world steel production expected to occur in Asia. Taking into account that carbon emissions by Chinese producers are far higher and more than double those of European producers, EUROFER arrives at the conclusion that the EU ETS leads to carbon leakage to non-ETS countries. The report notes that only about 30 percent of the world's steel producing countries have signed up to the obligations of the Kyoto agreement whereas 90 percent of all new capacity is being developed in the 70 percent not covered by a Kyoto obligation. The incentive for China, India, and other rapidly developing countries to join a cap-and-trade system is greatly diminished under these circumstances. Participating in a cap-and-trade treaty would mean affecting the economic growth that has occurred due to competitive advantages of not participating in a binding and targeted emissions reduction scheme. Key overseas competitors are reaping the economic benefits as production cuts in Europe due to carbon constraints begin to take effect (EUROFER 2009).

Evidence of the extent of carbon leakage from the EU to China is substantiated by various reports and predicted to be significant based on various projections of trade and energy intensity of production (i.e., GTAP-E models).[23] China has been expanding its market share in energy-intensive manufacturing, in particular in the iron, steel, and aluminum industries and Germany stands to be significantly affected by carbon leakage. As the most industrialized economy of the EU-15, Germany is heavily dependent on export and trade exposure of particularly its iron, steel, and aluminum industry is substantial. The EUROFER (2009) has studied the impact of emissions trading taking into account that steel is made in one of two ways: basic oxygen furnace (BOF) primary production and electric arc furnace (EAF) production involving recycling of scrap metal.[24] Nearly 100 percent of emissions in the EAF process are indirect electricity-related emissions, whereas 90 percent of BOF production is direct (i.e., process-related) emission and only 10 percent is indirect energy-related emissions. Products produced by the EAF process (recycling scrap metal) compete mostly in regional markets and are therefore able to pass on their production costs to their consumers. Conversely, the production of cold rolled flat steel using the BOF process is competing in global markets.

The EU aluminum industry is most threatened by the implementation of carbon constraints. Although excluded from participation during the first phase of the EU ETS, the aluminum industry is severely exposed to higher energy costs because of higher electricity prices and because the price of aluminum is set at a global exchange. Half of the EU's aluminum is produced by primary smelting and half by secondary smelting/recycling. The process of smelting consumes over 15 megawatt hour of electricity per tonne of aluminum, which immediately exposes the industry to cost increases due to energy price increases. As the industry is not able to pass on the costs of abatement to the customer, the EU is losing market share. The industry held 21 percent of world market share in 1982 but in 2010 the EU share of the global aluminum market was to 10 percent.[25]

A New Scenario for Climate Change Policy in a Globalizing World

Although various considerations play a role in location decisions, a factor of increasing importance noted by corporate investors and policymakers is differential environmental costs and regulatory burdens across national boundaries. Traditionally, environmental costs were external to the cost of production but under the conditions of the Kyoto protocol and the EU ETS this is no longer the case. Taking increased carbon costs and energy and electricity prices into account, the cost of production—in particular energy costs for energy-intensive industries—are substantially higher in the EU than in most parts of the world. Simultaneously, the institutionalization of a global free trade and investment regime under the WTO, the World Bank, and the IMF induces multinational corporations to relocate or shift production to countries where environmental regulations are less stringent and where no carbon constraints apply. Host economies encourage the export of high-carbon content products to developed countries' home markets and have set up free-trade zones or export platforms for that purpose. The combined effect has unintended consequences for climate change policy and is the driving force behind the increase in consumption of fossil fuels, increases in CO_2 emissions, and the development of a fossil fuel-based (or more specifically coal-based) infrastructure in developing countries. Despite the efforts to curb global CO_2 emissions through the implementation of the Kyoto Protocol, global carbon emissions have increased dramatically. While emissions have stabilized in some of the developed countries, international trade and foreign investment by multinational corporations has shifted the source of CO_2 emissions to developing countries (Schreuder 2009).

How to deal with and account for emissions embedded in products manufactured in nonabatement countries and exported to countries where carbon constraints are in effect remains an unresolved issue. Under the current international trade regime guided by the principles of the WTO, there is no compensation for the negative spillover effects of the shift in production from abatement to nonabatement countries. In fact, any measure that would compensate for the effect of the implementation of the Kyoto Protocol or the ETS in the context of the free-trade and investment regime that is currently in place would likely be interpreted as protectionism in disguise and would not be well received by the business community in the developing world. Whereas we might think of introducing import taxes (tariffs) on high carbon content goods and services or impose some other polluter pays principle on multinational corporations that relocate or shift production to noncarbon constrained parts of the world, in reality it would seem unlikely that any measures will be taken by the same institutions that established the neoliberal trade regime in the first place.

Derived from the same concern about unfair competition in a partial carbon-constrained global economy, we might also consider a more sector-specific approach to emissions trading. Different plants in a specific industrial sector in different parts of the world experience different cost structures and energy-intensive industries in Annex I countries where carbon constraints apply are most vulnerable to international trade if they cannot pass on the increased cost of production to their customers. Instead of accepting the business-as-usual response to competition and shift production to nonabatement countries, a particular sector or industry could decide to take action at the international level through global trade associations and pledge to achieve a sectorwide goal of reducing GHG emissions through sharing best practice standards and stimulating increased energy efficiency and technology transfer. Sector- or industry-specific initiatives could seek endorsement from national governments but because industry is not a party to the UNFCCC, such a scheme has little chance of success. Furthermore, such a solution might alienate developing nations and undermine the cooperation needed to move negotiations forward.

One alternative to the problems associated with the current climate change policy regime is to focus the frame of reference more specifically on the multinational corporation and ask the question who is to blame and who is to pay, or who stands to gain from the carbon embedded in imported goods manufactured abroad by multinational corporations. Is it the producer, the consumer, or the shareholder who profits from the investments made abroad for the purpose of avoiding carbon constraints (Schreuder 2009)? Full disclosure of ownership and full accounting of corporate activities throughout the production chain would be imperative under such a framework. In other words, at every step in the production process we would need to know how much CO_2 is added to the product and the national GHG pool. This amount would then be subtracted from the total amount of the country where multinational corporate production occurs and be added to the corporate home country's CO_2 budget or the country to which the product is shipped for consumption. Hypothetically, we could also attribute CO_2 emissions on a profit-rate basis. If a refinery in China operated by a foreign-owned multinational corporation generated 100 million tones of CO_2 per year and if 50 percent of the profits are returned to headquarters or to the multinational corporation's shareholders, then 50 mt of CO_2 emissions should be accounted for and subtracted from the host country's national emissions budget and added to the home country's CO_2 budget. All this would require far greater transparency than is presently the case and will likely be resisted by the corporate establishment. But unless we come to realize that global climate change policy requires global cooperation, we will not make much progress in global CO_2 reduction efforts. As part of reaching global consensus and cooperation, global corporations will have to restructure the way they operate in developing countries and their share in increase in global CO_2 emissions will have to be accounted for.

▪ **YDA SCHREUDER** is a professor of Geography and senior policy fellow in the Center for Energy and Environmental Policy at the University of Delaware. She is the author of *The Corporate Greenhouse: Climate Change Policy in a Globalizing World* (2009). She has also written several articles on the topic—published in *Bulletin of Science, Technology and Society*, and *Energy and Environment*; and has researched the implications of the UNFCCC climate change regime and the EU Emissions Trading Scheme on global shifts in production of energy intensive industries. An earlier version of this article was presented at the Nature Inc. Conference in The Hague, the Netherlands, in 2011.

▪ NOTES

1. Greenhouse gas is a gas that allows high-temperature solar radiation to enter the Earth atmosphere unhindered but blocks the lower-temperature reradiation of heat from the planetary surface causing the so-called greenhouse effect. Greenhouse gases include carbondioxide (CO_2), hydrochlorofluorocarbon (HCFC), hydrofluorcarbon (HFC), nitrousoxide (N_2O), methane (CH_4), ozone (O_3), perfluorocarbon (PFC), sulphurhexafluoride (SF_6), and water vapor. Of the greenhouse gases, CO_2 is the principal greenhouse gas responsible for global warming.

2. Parties to the UNFCCC are classified as: (1) Annex I countries, i.e. industrialized countries and economies in transition in East-Central Europe; and (2) non-Annex I countries, i.e. developing countries. Annex I countries that ratified the Protocol have committed to reduce their emission levels of greenhouse gasses to targets set mainly below their 1990 levels. Non-Annex I countries are not required to reduce emission levels but may participate in the Kyoto Protocol if they subscribe to the global emissions reduction efforts. As parties to the UNFCCC Kyoto Protocol they can participate in the Clean Development Mechanism. Setting no immediate restrictions under UNFCCC for developing countries serves three purposes: (1) it avoids restrictions on their development, because emissions are strongly linked to industrial capacity using fossil fuels; (2) developing countries can sell emissions credits through the Clean Development Mechanism (CDM) to Annex I countries that have difficulty meeting their emissions targets; and (3) developing countries can receive money and technologies for low-carbon investments from technologically advanced industrialized Annex I countries. Developing countries may volunteer to become Annex I countries when they are sufficiently developed. For further detail see the website for UNFCCC: http://unfccc.int/kyoto_protocol/items/2830.php/.

3. The Kyoto Protocol covers 194 countries including developing countries that subscribe to the global emissions reduction efforts.

4. The CDM is defined in Article 12 of the Kyoto Protocol, stating that Annex I countries can undertake emissions reduction projects in developing countries (non-Annex I) for Certified Emissions Reduction (CER) credits that can be used for compliance with the agreed to emissions reduction target on the part of industrialized Annex I countries.

5. Carbon leakage is usually expressed as a percentage of the CO_2 emissions increase that results from emissions increases in a nonabating (non-Annex I) country divided by the reduction of emissions by a country subject to the emissions-reduction target under the terms of the Kyoto Protocol.

6. See also World Bank, *World Development Report*, 67, quoted in The Ecologist, *Whose Common Future?* (1993: 111).

7. The coup against Allende had been backed by the CIA and was supported by the-then US secretary of state Henry Kissinger. Allende's successor, General Augusto Pinochet, called in a group of economic advisors who had been trained by Milton Friedman at the University of Chicago. They advised the general to transform the Chilean economy along free-market principles, privatizing public assets, opening up natural resources to private investors, and facilitating foreign direct investment. Export-led growth was to replace import substitution, which until then had been the dominant model for development in Latin America.

8. Friedrich August von Hayek (1899–1992) was best known for his defense of classical liberalism. His pioneering work in the theory and the analysis of the interdependence of economic, social, and institutional phenomena brought him great renown. He considered the efficient allocation of capital to be the most important factor leading to sustainable and optimal GDP growth, and warned of harm from monetary manipulation of interest rates. Milton Friedman (1912–2006) taught at the University of Chicago for more than three decades and is best known for his research on consumption analysis and monetary history and theory. As a leader of the Chicago School of Economics he influenced the research agenda of the economics profession in the United States in the second half of the twentieth century.

9. Energy-intensive industries include power generation, oil refineries, coke ovens, cement and concrete manufacturing, paper and pulp industries, glass and limestone, and steel and metals.

10. At present, China is the world's third largest net importer of oil behind the United States and Japan. Accordingly, carbon dioxide emissions from China and India have increased at an alarming rate, but, whereas the total amount of carbon emitted by China and India is approaching levels of the traditional industrialized countries, like Europe, Japan, and the United States, the per capita carbon emissions rates are still far behind those of the more mature industrialized economies (World Watch Institute 2006).

11. Under the Flexible Mechanism of the Kyoto Protocol, Annex I parties can contribute to their emissions targets by investing in emissions-reduction projects in other Annex I countries (Article 6, Joint Implementation), or by undertaking emissions reduction projects in developing countries (non-Annex I), defined in Article 12 as Clean Development Mechanism. Emissions Trading as defined in Article 17 of the Kyoto Protocol allows Annex I parties to acquire emissions credits from other Annex I parties and use them for compliance under the Kyoto Protocol. See www.unfccc.int/kyoto_protocol/items.

12. The first phase of the ETS allotted free allowances for target industries; the second phase auctioned allowances at a predetermined set price.

13. During the first phase, the aluminum industry was excluded from the ETS because it was fully exposed to global competition. However, because of high indirect costs due to high electricity prices—as utilities were not excluded from the ETS and windfall profits were charged—the industry was severely affected.

14. Windfall profits among energy providers occurred during the first phase of the ETS when the value of CO_2 allowances were passed on as opportunity costs even though the allowances were free and electricity may have been generated from noncarbon sources. The situation was enhanced by the fact that European power markets are not truly competitive.

15. GTAP is a global network of researchers and policymakers conducting quantitative analysis of international policy issues. It is coordinated by the Center for Global Trade Analysis at Purdue University's Department of Agricultural Economics.

16. Concern about the impact of the Kyoto Protocol on the US economy was clearly expressed in the Byrd-Hagel resolution in the Senate in 1997, which opposed the ratification of the Kyoto Protocol. Free trade would weaken the effectiveness of the Kyoto Protocol and emissions reduction schemes, according to US legislators and would bring harm to the US economy and to the global atmosphere.

17. EU-25 excludes Romania and Bulgaria, which were admitted to the EU in 2007 and only participated in the EU-ETS since 2008.

18. Following the agreement among EU member states and the European Parliament, the EU ETS Directive was significantly revised, as part of the EU 2020 Climate and Energy Package in December 2008, which established a 20 percent emissions reduction target by 2020 based on 1990 levels.

19. Under the Clean Air Act, the Environmental Protection Agency has been authorized to control CO_2 emissions for new power plants. Any new plant built in the United States cannot emit more than 1,000 pounds of CO_2 per megawatt hour. The vast majority of modern natural-gas plants meet that standard but conventional coal plants average upward of 1,800 pounds per megawatt hour. This effectively means that it will be impossible to build any new coal-fired power plant in the United States that cannot capture and store its own carbon emissions. For the time being this means that there is a moratorium on all new coal-powered plants.

20. Setting ambitious unilateral EU emissions reduction targets while mitigation efforts in third countries are limited should be discouraged according to the European Alliance of Energy Intensive Industries March 7, 2012 statement in response to the EU Commission's roadmap to a competitive low-carbon economy by 2050. http://www.cembureau.eu/sites/default/files/documents/2012-03-07%20Alliance%20statement%20low-carbon%20economy%20roadmap%202050.pdf (accessed 15 May 2012).

21. Reuters Business News, June 6, 2007 (accessed 9 June 2009).

22. A study conducted by researchers from the US National Center for Atmospheric Research (NCAR) confirmed this and calculated that between 1997 and 2005 higher levels of Chinese exports to the United States increased total carbon dioxide emissions by some 720 million tonnes (Sato et al. 2007)

23. See the previous section on the EU Emissions Trading Scheme and Carbon Leakage.

24. The European Confederation of Iron and Steel Industries (EUROFER) website: http://www.eurofer.org.

25. See European Aluminium Association: http://www.alueurope.eu/about-aluminium/facts-and-figures (accessed 12 September 2012).

■ REFERENCES

Alliance of Energy Intensive Industries. January 2004. "Energy Intensive Industries call upon EU Decision Makers to Pay More Attention to the Impact of Emissions Trading upon Their Competitiveness." See www.eula.be (accessed 5 May 2006).

Alliance of Energy Intensive Industries. November 2005. "The Impact of EU Emissions Trading Scheme (ETS) on Power Prices." http://www.cembureau.beCem_warehouse/ALLIANCE%20ETS%20AND%20POWER%20PRICES.PDF (accessed 2 June 2006).

American Iron and Steel Institute (AISI). 2006. "Environmental Aspects of Global Trade in Steel: The North American Steel Industry Perspective." http:www.steel.org (accessed 4 June 2008).

Babiker, Mustafa H. 2005. "Climate Change Policy, Market Structure and Carbon Leakage." *Journal of International Economics* 65, no. 2: 421–445.

British Petroleum (BP). 2008. *Statistical Review of World Energy.* http://www.bp.com/productlanding.do?categoryId=6929&contentId=7044622 (accessed 12 June 2009).

British Petroleum. 2011. *Energy Outlook 2030.* London. http://www.bp.com/liveassets/bp_internet/spain/STAGING/home_assets/downloads_pdfs/e/energy_outlook_2030.pdf (accessed 26 December 2011).

Chomsky, Noam. 1999. *Profit over People: Neoliberalism and Global Order.* New York: Seven Stories Press.

CRU News. 2006. "Higher Power costs Driving Aluminum Smelter Closures in Europe and the USA." 30 March. http://cruonline.crugroup.com/Aluminium/tabid/117/Default.aspx (accessed 15 May 2012).

Daly, Herman. 1996. *Beyond Growth: The Economics of Sustainable Development.* Boston: Beacon Press.

The Ecologist. 1993. *Whose Common Future? Reclaiming the Commons.* Philadelphia: New Society Press; London: Earthscan.

Econsense. 2007. Forum for Sustainable Development of German Business. www.climate-policy-map.econsense.de/datasources.html.

Ellerman, Denny, and Paul L. Joskow. 2008. *The European Union's Emissions Trading System in Perspective.* Washington, DC: Pew Center on Global Climate Change.

Energy Information Administration (EIA). 2006. "International Energy Outlook 2006." http://www.china-profile.com/data/fig_co2-emissions_2.htm (accessed 26 December 2011).

European Alliance of Energy Intensive Industries. 2012. http://www.cembureau.eu/sites/default/files/documents/2012-03-07%20Alliance%20statement%20low-carbon%20economy%20roadmap%202050.pdf (accessed 15 May 2012) .

European Aluminum Association. 2011. "Addressing the Future of the Aluminum Industry in Europe." www.eesc.europe.eu/resources/docs/schrynmakers (accessed 15 May 2012).

European Association for Coal and Lignite (EURACOAL). 2009. "Market Report." http://www.euracoal.be/pages/home.php?idpage=1 (accessed 8 June 2009).

European Confederation of Iron and Steel Industries (EUROFER). 2009. *Combating Climate Change.* http://www.eurofer.org/index.php/eng/News-Publications/Publications (accessed 15 June 2009).

Giddens, Anthony. 2009. *The Politics of Climate Change.* Cambridge: Polity Press.

Global Carbon Project. 2010. "Carbon Budget 2010." www.globalcarbonproject.org (accessed 26 December 2011).

Hamasaki, Hiroshi. 2007. "Carbon Leakage and a Post-Kyoto Framework." Fujitsu Research Institute Working Paper #287, Tokyo.

Harvey, David. 2005. *A Brief History of Neoliberalism.* Oxford and New York: Oxford University Press.

Harvey, David. 2006. *Spaces of Global Capitalism: Towards a Theory of Uneven Geographical Development.* New York: Verso.

Hillebrand, Bernard et al. 2002. *CO_2 Emission Trading Put to the Test.* Munster: LIT Verlag.

Intergovernmental Panel on Climate Change (IPCC). 2007. "Climate Change 2007: The Physical Science Basis; 'Summary for Policy Makers.'" http://www.ipcc.ch/publications_and_data/ar4/wg1/en/contents.html.

International Energy Agency (IEA). 2008. "World Energy Outlook." www.worldenergyoutlook.org (accessed 17 June 2009).

International Energy Agency. 2012. "Tracking Clean Energy Progress." www.iea.org/papers/2012 (accessed 25 April 2012).

Organization for Economic Cooperation and Development (OECD). 2006. "International Investment Perspectives." http://www.oecd.org/document/37/0,3746,en_2649_33763_37449253_1_1_1_1,00.html (accessed January 2008).

Klein, Naomi. 2008. *The Shock Doctrine: The Rise of Disaster Capitalism.* New York: Metropolitan Books.

Kuik, O., and R. Gerlagh. 2003. "Trade Liberalization and Carbon Leakage." *Energy Journal* 24, no. 3: 97–120.

Koch, Max. 2011. *Capitalism and Climate Change: Theoretical Discussion, Historical Development, and Policy Responses.* New York: Palgrave Macmillan.

Li, Minqi. 2008. *The Rise of China and the Demise of the Capitalist World Economy.* New York: Monthly Review Press.

Newell, Peter, and Matthew Paterson. 2010. *Climate Capitalism: Global Warming and the Transformation of the Global Economy.* Cambridge: Cambridge University Press.

O'Connor, Martin, ed. 1994. *Is Capitalism Sustainable? Political Economy and the Politics of Ecology.* New York: Guilford Press.

Peet, Richard. 2007. *Geography of Power: Making Global Economic Policy.* London: Zed Books.

Peters, Glen P. et al. 2011. "Growth in Emissions Transfers via International Trade from 1990–2008." In *Proceedings of the National Academy of Sciences.* Oslo: Center for International Climate and Environmental Research. http://www.cicero.uio.no/webnews/index_e.aspx?id=11540 (accessed 26 December 2011).

Porritt, Jonathan. 2005. *Capitalism as If the World Matters.* London: Earthscan.

Sachs, Wolfgang. 1999. *Planet Dialectics: Explorations in Environment and Development.* London: Zed Books.

Sachs, Wolfgang, ed. 2002. *The Jo'burg Memo: Fairness in a Fragile World. Memorandum for the World Summit on Sustainable Development.* Berlin: Heinrich Böll Foundation.

Sato, Misato et al. 2007. "Differentiation and Dynamics of Competitiveness Impacts from the EU ETS." 12 April. www.electricitypolicy.oth.uk/TSEC/2/differentiationdynamics.pdf. (accessed 12 January 2008)

Schreuder, Yda 2009. *The Corporate Greenhouse: Climate Change Policy in a Globalizing World.* New York and London: Zed Books.

Sen, Amartya. 1999. *Development as Freedom.* Oxford: Oxford University Press.

Sijms, J.P.M. et al. 2005. *Spillovers of Climate Policy: An Assessment of the Incidence of Carbon Leakage and Induced Technological Change due to CO_2 Abatement Measures.* The Hague: Netherlands Research Programme on Climate Change, Rijksinstituut voor Volksgezondheid en Milieu.

Slater, D.R., and Taylor P., eds. 1999. *The American Century.* Oxford: Blackwell.

Stern, Nicholas. 2006. *The Economics of Climate Change: The Stern Review.* Cambridge: Cambridge University Press.

Stiglitz, Joseph E. 2002. *Globalization and it Discontents.* New York: Norton.

Taylor, M. Scott, and Brian R. Copeland. 2005. "Free Trade and Global Warming: A Trade Theory View of the Kyoto Protocol." *Journal of Environmental Economics and Management* 49, no. 2: 205–234.

United Nations Framework Convention on Climate Change (UNFCCC). Kyoto Protocol: http://unfccc .int/kyoto_protocol/items/2830.php/ (accessed 5 March 2012)

United States Department of Energy. 2006. "Country Analysis Briefs, China." http://www.geni.org/ globalenergy/library/national_energy_grid/china/china_country_analysis_brief.shtml (accessed January 2008).

Valdez, Juan P. 1995. *Pinochet's Economists: The Chicago School in Chili.* Cambridge: Cambridge University Press.

Wallerstein, Immanuel. 2000. *The Essential Wallerstein.* New York: New Press.

Weinreich, Dirk. 2008. "How to Address Carbon Leakage in the EU ETS." Federal Ministry for the Environment, Nature Conservation and Nuclear Safety. http://ec.europa.eu/environment/climat/ emission/pdf/080411/de_carbon_leakage.pdf (accessed 25 May 2009).

Weyant John P., ed. 1999. "The Costs of the Kyoto Protocol: A Multi-Model Evaluation." *Energy Journal* (Special Issue of the International Association for Energy Economics). http://www.iaee.org/en/ publications/kyoto.aspx

Williamson, John. 1990. *Latin America: How Much Has Happened.* Washington, DC: Institute of International Economics.

World Commission on Environment and Development (WCED). 1987. *Our Common Future.* Oxford and New York: Oxford University Press.

World Health Organization (WHO). 2005. *Ecosystems and Human Wellbeing: A Report of the Millennium Ecosystem Assessment.* Geneva. http://www.maweb.org/documents/document.356.aspx.pdf

World Watch Institute. 2006. "State of the World 2006." http://www.worldwatch.org/bookstore/ publication/state-world-2006-special-focus-china-and-india (accessed 25 March 2007).

Zhang, Si, and Pearce, Robert. 2012. *Multinationals in China: Business Strategy, Technology and Economic Development.* New York: Palgrave Macmillan.

BOOK REVIEWS

BUTTON, Gregory, *Disaster Culture: Knowledge and Uncertainty in the Wake of Human and Environmental Catastrophe,* 2010, 311 pp. Walnut Creek, CA: Left Coast Press. ISBN 978-1-598-74389-0.

With more than 30 years of research to draw from, internationally recognized expert in the anthropology of disaster, Gregory Button, offers this analysis of the ways in which uncertainty is culturally produced in the wake of calamity. He contends that while analyses of risk—as assessments of probability and possibility—have been thoroughly addressed in recent literature, the role of uncertainty has been long trivialized as irrational concerns too elusive to yield useful analytical data, even though senses of uncertainty are fundamental to notions of risk. Button uses the powerful anthropological strategy of comparative analysis across seven ethnographic studies of disaster to analyze discourses of uncertainty as they are produced socially and institutionally. He presents his research to argue that experiences and responses to disaster are shaped by both diffuse forms of informational uncertainty and institutions seeking to manipulate an already vague environment to their advantage.

With regard to the socially constructed on-the-ground experience of individuals and communities, the post-disaster environment is filled with conflicting information about severity and risk, often the product of media and various agencies' swift, yet uncoordinated, public communications. Button argues that as people affected by disaster try to understand their situation, this onslaught of disjointed information is received as confusing and conflicting. This leads to a general distrust of disaster-related information whether from national news media or scientific agencies.

Throughout his examples, Button is careful to show how uncertainty about the validity of information available leads to both individual and community distress that further exacerbates understanding and effective decision making. However, the thrust of Button's analysis focuses on how agencies and institutions take advantage of and manipulate this precarious informational environment to politicize local and national discourses of disaster in ways that benefit those tasked with managing disaster response. He argues that the production of uncertainty is an intentional ideological tactic—that the "control of public discourse, as well as the attempt to control the social production of meaning, is an attempt to define reality" (p. 16). And he shows how thoroughly agencies charged with disaster management are able to manipulate this uncertainty to their advantage.

Button demonstrates his arguments through his selection of ethnographic sketches or "case studies" of varied forms of disaster. Although not full case studies per se, each is crafted to make an analytic point and successfully does so. From the Exxon-Valdez oil spill to Three Mile Island, and from the 9/11 terrorist attacks to Hurricane Katrina, Button shows how the manipulation of information, either by manufacturing it outright or withholding it, is deployed to increase uncertainty. His analyses also consistently demonstrate how the imme-

Environment and Society: Advances in Research 3 (2012): 123–150 © Berghahn Books
doi:10.3167/ares.2012.030108

diacy of media-disseminated information and the more necessarily methodical pace of scientific response compound this environment of uncertainty—often unintentionally in the communications of victims making sense of their postdisaster world, and often quite intentionally on the behalf of the economically and politically motivated corporations and agencies involved. The remainder of the text offers more in-depth analyses of the case study data and expands on the book's major themes: how accounts of disaster are mediated; how knowledge becomes contested and segregated; and how uncertainty at all levels of disaster experience is culturally produced.

As with other works by Gregory Button, this book is well written, accessible, and sophisticated in its delivery of the larger message. At moments the social injustices documented by Button are deeply distressing and even heart wrenching. His ethnographies transport the reader from instances when bureaucratic impediments threatened to stop his own research even as fabricated information was being disseminated, to the kitchen table in a Federal Emergency Management Agency (FEMA) trailer where a family's fears about environmentally induced sicknesses are powerfully recounted. But perhaps most impressive is the author's ability to address the structural issues at hand while maintaining a sense of people's on-the-ground experience and not losing sight of the humanity of disaster.

Button closes his book with the addition of a final chapter based on the BP oil spill in the Gulf of Mexico that was unfolding as the book was being completed. This last case study impresses on the reader both the timeliness of this work and Button's valuable contributions to the field of disaster research. Indeed, looking at disaster though a longer-term lens shows how the ideological tactics of uncertainty have developed and continue to shape experiences of disaster today. Button's work demonstrates that, while disaster is most often imagined as a moment of exception, the processes of response and meaning making

have become routinized and normalized such that uncertainty is broadly perceived of as an unquestioned "natural" postdisaster state rather than a socially and institutionally produced cultural experience. By deconstructing this process, anthropology can fruitfully demystify experiences of disaster and contribute to effecting change in the ways that information is produced, withheld, and used to limit victims' decision making and avenues of recourse.

SherriLynn Colby-Bottel
California State University, Fresno

FALASCA-ZAMPONI, Simonetta, *Waste and Consumption: Capitalism, the Environment, and the Life of Things*, 65 pp. New York: Routledge, 2010. ISBN 978-0-4158-9210-0.

"How many places will we need to build and then forget, I wonder?" This is a question that sociologist Simonetta Falasca-Zamponi asks in her book, *Waste and Consumption*, which explores the environmental impacts of capitalist consumption. Here she refers to the legacy of radioactive waste "we" are leaving behind, but would sooner ignore. The question remains, like the book as a whole, more of a provocation than an invitation to engage in more detailed analysis. This is partly out of necessity. As a contribution to the Routledge Social Issues Collection, *Waste and Consumption* is a very compact text, meant to serve as an accessible overview of sociological approaches toward contemporary public concerns. Its main audience is "instructors teaching a wide range of courses in the social sciences" and each chapter is followed by several questions designed to generate discussion among students and work groups. Perhaps due to this format, the book has its limitations.

Falasca-Zamponi explains the purpose of the book as an attempt to diagnose the convictions of climate skeptics, particularly the bloggers, corporations, and libertarians that contest the climate consensus and challenge those who would further regulate their economic behavior. For Falasca-Zamponi, climate skeptics are united in their "defense of consumption" (p. 3), which leads them to value the right to consume above responsibility for the environment. Her lack of references to any social science analysis of these debates, especially with regard to the politically and historically mediated relationship between climate skepticism and science in the United States, is symptomatic of the book's general style, which suffers from an inconsistent engagement with the scholarly literature generally. This is particularly the case, unfortunately, with the book's twin themes of waste and consumption.

Falasca-Zamponi's initial introduction to consumption is adequate as a historical treatise, but strangely fails to draw on the vast field of consumption studies. It is as if consumption were primarily a matter of economic history, of acquiring utilities, rather than a process with social and symbolic significance. These passages are not uninformative, but they do not provide a sociological introduction to the "life of things" within capitalism, or explain why some should feel so attached to the "right to consume." The book's discussion of waste suffers from the opposite problem. Whereas consumption is largely framed in terms of economic history, waste is interpreted through the familiar synchronic analysis of Mary Douglas's *Purity and Danger* (1966); it is therefore cast as a seemingly cross-cultural and transhistorical tendency to categorize certain people and things as "out of place." A more detailed account might have consulted Marxian histories of such hierarchies and their relationship to the emergence of industrial capitalism (e.g., Stallybrass and White 1986). Even more relevant would have been sociologist Martin O'Brien's recent book, *A Crisis in Rubbish?* (2008), which explores the central role of waste in the birth of industrialization and its changing nature over the intervening centuries.

Where waste is given a history, it is decidedly uneven. As Falasca-Zamponi explains

the risks of the primary methods of disposal–landfill and incinerators–technical improvements to the latter receive far more attention and are described approvingly, with official documents from the Danish government used, curiously, as support (pp. 20–21). Discussion of global waste trades similarly do not review any significant changes since the 1980s, especially the Basel Convention, which officially banned the shipment of waste from countries of the Organization for Economic Cooperation and Development (OECD) to non-OECD countries, with some success, or China's rapid urbanization and its effect on the global market for recycled plastics and metals (see Clapp 2001; Alexander and Reno 2012).

Falasca-Zamponi's concludes with an appeal to Bataille's theory of expenditure, which is interesting if underdeveloped. In particular, it is unclear what "a broader affirmation of the individual right to pleasure and aesthetic enjoyment" (p. 48) would add to the debate over climate change per se. Although, one could imagine that if this included appreciation of (and a sense of responsibility for) the environment, or things themselves, then it might hold more radical potential. The author begins to explore some of these themes in a fascinating discussion of fascist Italy and Mussolini's paradoxical attempt to sever consumption from capitalism. Along with a case study in Seveso, Italy, involving public exposure to toxic emissions, these are among the best passages in the book. But these promising threads are not satisfactorily drawn together. Further, readers are not informed where they could read more about the possibilities associated with an aesthetic politics of waste, in Hawkins (2006) or Neville and Villeneuve (2002), or where some have offered critiques of Bataille's relatively static theories of waste in favor of more dynamic approaches (e.g., Frow 2003).

The neglect of so much relevant material does not impede the book's agenda, which is to prove that consumption is morally problematic and use waste as a way of pointing this out. For this reason, waste comes to mean not only actual trash, but also air pollution, toxicity in food, carbon emissions, nitrates in the water supply; in short, waste stands in for environmental destruction more broadly. I share the author's convictions, but I fear that a number of these claims are supported with too little evidence and do not offer clear pathways toward further scholarly engagement with environmental issues. In another book on waste that is not referenced by the author, sociologist Zsuzsa Gille (2007) criticizes "end of pipe" analyses of environmental problems that fail to account for whole "waste regimes." *Waste and Consumption*, to its credit, attempts such a broad-based, historically informed analysis, but as an overview of scholarship on these issues it is lacking. Falasca-Zamponi has written an impassioned defense of the powers and dangers of consumption, but students may need additional guidance if they are meant to delve deeper into the social study of the environment.

Joshua Reno
Department of Anthropology
Binghamton University

References

Alexander, Catherine and Joshua Reno, eds. 2012. *Economies of Recycling: The Global Transformation of Materials, Values and Social Relations*. London and New York: Zed Books.

Clapp, Jennifer. 2001. *Toxic Exports: The Transfer of Hazardous Wastes from Rich to Poor Countries*. Ithaca, NY: Cornell University Press.

Douglas, Mary. 1966. *Purity and Danger*. London: Routledge and Kegan Paul.

Frow, John. 2003. "Invidious Distinction: Waste, Difference, and Classy Stuff." Pp. 25–38 in *Culture and Waste: The Creation and Destruction of Value*, ed. G. Hawking and S. Muecke. Oxford: Rowman and Littlefield.

Gille, Zsuzsa. 2007. *From the Cult of Waste to the Trash Heap of History*. Bloomington: Indiana University Press.

Hawkins, Gay. 2006. *The Ethics of Waste*. Oxford: Rowman and Littlefield.

Neville, Brian, and Johanne Villeneuve. 2002. *Waste-Site Stories: The Recycling of Memory*. Albany: State University of New York Press.

O'Brien, Martin. 2008. *A Crisis in Rubbish?* London and New York: Routledge.

Stallybrass, Peter, and Allon White. 1986. *The Politics and Poetics of Transgression*. London and Cambridge: Taylor and Francis.

FIJN, Natasha, *Living with Herds: Human-Animal Coexistence in Mongolia,* 274 pp. Cambridge: Cambridge University Press, 2011. ISBN 978-1-107-00090-2.

I am confident that *Living with Herds* will prove an authoritative handbook for the anthropology of Mongolian herding practices. Already, it is a welcome addition to the limited English corpus. It provides the next link in historical continuity of the existing literature: Herbert Vreeland (1954) described the modern situation on the eve of revolution, Tserendash Namkhainyambuu (2000) covered a moment in socialism, and now Fijn brings us into the post-socialist twenty-first century. Fijn does not set out to compare the present to the past, but readers familiar with the literature will marvel at the transformation: gone is the puppet monarchy, gone are the powerful Buddhist temples and their influence, gone is the effect of military conscription on the household labor pool, and gone are the socialist herding collectives and their ideological rationalism.

These transformations lurk in the background of many of Fijn's descriptions of herding and household management practices, and could benefit from being made explicit. As it stands, many questions linger: Did the end of communism enable the return of dormant presocialist practices, or were these reinvented? Has Soviet technology, heavy-handedly introduced, left any social traces now that it is gone? And how have new joint-stock herding enterprises and market competition affected herding life and the animals? There is a need for more ethnography and comparative historical research to answer these questions.

The decision to mostly ignore the historical and political context may be related to the book's broader ambitions of contributing to an ongoing discussion within anthropology about human-animal relationships, especially those in which animals are considered as social participants. As such, the present Mongolian case is used as an example of possibility for such relationships, for which the sociohistorical contingencies of custom may seem less important. How well does this approach work?

We find several moments of classic ethnographic synthesis, in particular the wonderfully clear and illustrative diagrams and tables that relate symbolic systems to time, space, and the contingencies of herding practice. Rich ethnographic vignettes underline the role of nature and animals. On the one hand, there are the material requirements for raising and caring for each species from birth to death according to their adaptability to climate and geography, as well as the diverse character of individual animals. On the other hand, there are the human cultural and economic needs for food, medicine, trade, cash, cashmere, transportation, and ritual materials, as well as religious and emotional dependencies on the animals.

Even without reading her analysis, this framework of mutual dependency would suggest a social structure in which animals are active participants. Beyond the web of consequence that ties the animals to "their" humans, animals respond as both individuals and groups to a changing social milieu. Fijn does a good job at drawing out a kind of animal's perspective on Mongolian culture, stopping short of anthropomorphizing them. She seems to learn how to walk this delicate line from her Mongolian informants, who are attentive to and respectful of their animals precisely as fellow social actors. Indeed, the categories of her analysis find considerable overlap with Mongolian native categories. Perhaps too comfortable an overlap.

This leads to a few blind spots. Though she describes the symbolism with regard to animals and their bodies (for example, the structure of a the *khot ail* is analogous to that of a herd animal's body), she fails to elaborate on the fact that much of this discourse indexes *dead* and *dismembered* animals, particularly their meat, fat, organs, bones, sinews, and so on. That animals are *food*, but humans *are not*, should be central to a mutual social structure. However, activities such as slaughter, dismemberment, and food preparation are underemphasized in the narrative. From an ethnographic fieldworker's perspective, killing is indeed a quick and small-seeming event as compared to the long hours and months spent herding, but one cannot ignore its deep cultural significance. And what is the animal's perspective on being slaughtered and cooked, as well as milked, sheared, bled? Fijn writes: "When I asked herders what happened to the animals after they died, I was met with silence; it would have been too awkward to pursue my question any further" (p. 226). Animals cannot answer any questions, dead or alive, yet the analysis of their role continues. There is a methodological awkwardness here that could benefit from explication.

Finally, it is unsurprising that Fijn disagrees with Timothy Ingold's conclusions about the social structure of herding. Whereas he emphasizes "domination," she sees more "mutuality." She has idealized the situation: mutuality and symbiosis obviously go very far, but they eventually reach the limits of both obstinate ideology and severe material necessities. Humans are cruel to animals, just as surely as they are cruel to each other, and to themselves. Fijn paints a convincing picture of ideas about mutuality, but does not follow through on the implications of her own evidence of domination. For example, she perceptively points out the subtle ways by which herders control the behavior of animals while still encouraging their free-range behavior, but does not investigate just what Mongolians think and feel about imposing such severe limitations.

History wants to intervene. How Mongolians interact with animals cannot be so easily separated from how they interact with fellow humans, a field that was severely affected most recently by decades of socialist state intervention: the bloody dismantling of Buddhism, kinship restructuring, Soviet colonialism, ethnopolitical maneuvering, and forced agrarian policy. Violence looms: tools and methods used to herd animals were all too easily used against humans. And this, too, has echoes that have been reverberating since prehistory. As categories, "domination" and "mutuality" are anything but immune to hypocrisy, cynicism, and blatant abuse. Ingold not only underscores this, but grounds his argument in history, indeed in historiography as a mode of domination.

But one book, and one span of research, cannot be expected to cover all things. Indeed, there is much to appreciate in Fijn's insistence on confining her analysis to the scope of her data. Her cautious, step-by-step syntheses are reassuring, and her straightforward, nonsensationalist tone, where others might be tempted to shrill, righteous screeds, is welcome. Confined as it is, the scope of her research is impressively expansive. She succeeds in adding much needed depth to our understanding of contemporary herding life in Mongolia, and has contributed to an important debate about the social agency of animals.

Tal Liron
Department of Anthropology
University of Chicago

References

Ingold, Timothy. 1994. "From Trust to Domination: An Alternative History of Human-Animal Relations." In *Animals and Human Society: Changing Perspectives,* ed. A. Manning and J. Serpell. Pp. 61–76. New York: Routledge.

Namkhainyambuu, Tserendash. 2000. *Bounty from the Sheep: Autobiography of a Herdsman.* Cambridge, UK: White Horse Press.

Vreeland, Herbert H. 1954. *Mongol Community and Kinship Structure.* New Haven, CT: Human Relations Area Files.

GUNERATNE, Arjun, ed., *Culture and the Environment in the Himalaya,* 256 pp. New York: Routledge, 2010. ISBN 978-0-4157-7883-1.

Few areas of the world are blessed with as dramatic a geography as the Himalayas. Its distinctive environment has defined the Himalayas as a region and helped shape the cultural practices, political identities, and economic possibilities of those who live there. Despite this fact, anthropologists have not devoted much attention to exploring how those who live in the Himalayas conceive of the environment, or how their cultural practices are shaped by it. As its title suggests, *Culture and Environment in the Himalaya* attempts to remedy this lack. Its editor, Arjun Guneratne, describes the aim of the volume to be the exploration of "[w]hat environment means to people who live and work in the Himalaya, and the impacts of those understandings … for policy-making in development work" (p. 1). This is an ambitious agenda, and the ten contributors to the volume pursue it in a number of ways.

Several tackle the theme head on by exploring how particular groups living in Nepal and northern India conceive of, and interact with, their environment. Andrew Russell examines, for example, the deeply ambivalent, even sometimes hostile, conception of the forest held by members of the Yakkha community in East Nepal. In a fascinating chapter, Jane Fortier examines how the Raute, a forest-dwelling group who live in western Nepal, subvert statist and Brahminical discourses that constitute them as the wild and backward Other by conceiving of themselves as subjects of the forest rather than subjects of the state. T.B. Subba explores how Limbu conceptions of their physical environment are intertwined with—and, Subba argues, inseparable from—their conceptions of the spiritual world.

Other contributors pursue the theme of culture and environment in the Himalayas by examining the impact of scientific or developmentalist discourses on local practices and conceptions of the natural world. Mary Cameron explores, for example, the efforts by the Nepalese state to regulate and rationalize the practice of Ayurveda in Nepal. She argues that the developmentalist state's emphasis on the formal medical education of Ayurvedic practitioners in fact impedes its effective practice. John J. Metz examines how the environmental crisis narrative (what he calls the "Theory of Himalayan Environmental Degradation" or THED) that dominated scientific and developmentalist discourses about the environment in Nepal throughout much of the 1970s and 1980s was subsequently undermined by evidence suggesting that processes of environmental change in the Himalayas were much more complex, and Himalayan peoples much more responsive to changing conditions, than researchers originally assumed. In a rather dense chapter, Ben Campbell examines how Tamang communities living in Langtang National Park know and narrate the natural world. He argues that the Tamang way of knowing challenges the nature/culture binary that is at the heart of developmentalist and conservationist discourses in Nepal. Safia Aggarwal examines how village *panchayats* attempt to mobilize local religious beliefs to conservationist ends by placing local forests under divine protection—and the complex consequences of doing so.

A final set of contributors examines the relationship between culture and environment by foregrounding a theme that is implicit in all the chapters: namely, how environmental discourses shape and are shaped by politics. Thus, Emma Mawdlsey examines how the Hindu nationalist group, the Vishwa Hindu Parishad, attempted to mobilize environmentalist opposition to the Tehri Dam project in Uttarkhand, India, to promote an anti-Muslim ideology. Anne Rademacher examines how a particular conservation discourse—what she terms the Bagmati Civilization narrative—linked a political critique of Nepalese government and society to an environmental critique of the state of the Bagmati River in Kathmandu, and the role it played as a result in struggles to reimagine political identity and community in Nepal during a period of profound upheaval. Finally, Andrea Nightingale examines how differently positioned groups in Mugu district in northwestern Nepal experienced changes in the management of the local forest over the course of the twentieth century, and how their historical narratives of the forest in turn index ongoing struggles over the access and control of forest resources.

As may be expected from a collected volume of papers—many of which were originally presented at a conference at Macalester College in 2004—the style and preoccupations of the chapters vary widely, from the focused ethnographic accounts of environmental practice presented in some of the earlier chapters to the more narrativistic and nation-state centered accounts of the politics of the environment presented in the later chapters. This is both a weakness and a strength of the book. For readers who are not already familiar with the political geography of the region, some of the early chapters in particular may prove tough going, because contributors assume a certain amount of background knowledge. Conversely, the diversity of the topics presented makes the volume a rewarding read for those who are willing to persevere, and interesting continuities emerge over the course of the book. Chief among these is the attempt by many of the contributors to develop a theoretical approach to questions of environment and culture that is neither overly positivist (like the crisis of environment narrative that Metz and others critique) nor overly idealist in its faith in infinite cultural adaptivity. In most cases, what this produces are subtle and historically attentive ethnographic accounts of the dynamic relationship between political, cultural, and ecological processes that are very enjoyable to read and a testament to the vitality of environmental anthropology in the region.

I have some quibbles. Few of the authors pay much attention to how large-scale economic and political processes, such as democratization (in Nepal) and industrialization generally, have affected the groups and practices they study. Nonetheless, we get hints throughout the book that such processes have had, and will continue to have, a considerable impact on ecological conditions and practices throughout the region. Similarly, although Subba and other authors note the profound effect that environmental degradation is having both on ecological practice and identity-formation, none of the chapters take up this theme explicitly, leaving the reader somewhat in the dark about the significance and larger consequence of the environmental changes that have taken place over the past fifty years. This may reflect the contributors' justified wariness of the unreflexive positivism that characterized the now much-critiqued THED. One can, however, remain sensitive to the complexity of the social and material worlds—and committed to a reflexive social science—and nonetheless engage in the kind of large-scale analysis that the volume lacks.

On the whole, however, this is a minor problem. More than anything, it points to the richness and complexity of the questions posed and raised by the volume. *Culture and Environment in the Himalaya* may not therefore be the last word on the question of the relationship between culture and the environment in the Himalayas. It nonetheless makes clear why the topic is worth examining, not only for what it reveals about the relationship between different groups and the environment but also because of the perspective it sheds on the region as a whole. As such, it represents a welcome and rather ambitious addition to Himalayan ethnography that should be of interest not only to environmental anthropologists but to Himalayanists generally.

Genevieve Lakier
Department of Anthropology
University of Chicago

HASTRUP, Frida. *Weathering the World: Recovery in the Wake of the Tsunami in a Tamil Fishing Village,* 158 pp. New York: Berghahn Books, 2011. ISBN 978-0-85745-199-6.

Weathering the World is the latest volume in the series Studies in Environmental Anthropology and Ethnobiology. Despite its publication within a domain-specific series, Frida Hastrup's book should rightly be recognized as a unique contribution to theoretical bridging between the traditionally separate orientations of inquiry labeled as the "anthropology of disasters" and "environmental anthropology." Supported and contextualized in rich chapters of intimate ethnography, the true strength and contribution of the book is found in Hastrup's reorientation of traditional theoretical constructs and modes of inquiry commonly employed in both environmental and disaster discourses.

Hastrup's research is based on ten months of anthropological fieldwork, spanned over a period of three years (2005–2008), in the small fishing village of Tharangambadi, off the eastern coast of Tamil Nadu, in southern India. Her approach is largely qualitative and theory building, with the methods section relegated to a brief mention of her primary approach to ethnographic engagement—walking through villages conducting loosely structured interviews. Secondary to these encounters, the author engaged with local and foreign aid workers, nurses, clergy, shop owners, and others. Regrettably, Hastrup does not report the number of villagers interviewed, or her sampling approach, bringing into question her ability to extrapolate her findings beyond her sample to the village itself. It would have been nice to see Hastrup's strong qualitative insights supported by figures grounded in, at the very least, her presumably purposive sample. For example, when Hastrup notes that "the *majority* of my interlocutors were women" (p. 33; emphasis added), the reader is left to wonder if this qualitative categorization represents 51 percent of her interlocutors, two-thirds, or 95 percent. Whatever limits to generalizability her sampling approach imposes, the absence of its details does not necessarily detract from the book's theoretical contributions. Although Hastrup does not articulate a specific research design, this reviewer cautiously concludes that her approach was likely that of a case study, which is entirely appropriate for theory building, testing, and refinement (Yin 2009)—which, as noted, is the principal contribution of the book.

Building on Susanna Hoffman's (1999) insight that much disaster research fixates on the notion of *change* (adaptation), Hastrup frames her study in the context of *recovery*. The very title of the book points to this central theme: that the tsunami was experienced not as a singular temporal event that occurred in the past, and from which the villagers had to adapt, but as an event from which the villagers continue to employ strategies of recovery, in an iterative and ongoing process of weathering the world. Hastrup's superb writing provides ethnographic support for this point of departure, illustrated in her treatment of the village fishermen's extension of traditional weather categories to include explanations for the tsunami, irregular rainfall patterns, and other overarching climatic changes.

Refreshingly, attention is duly paid to various material aspects of the disaster in an attempt to balance an anthropology of disaster which, according to Hastrup, "has paid too scant attention to the concrete physicality of catastrophes by emphasising the role of the social construction and politics of disaster in post-calamity settings" (p. 26). Of particular interest is her treatment of the curious and widespread habit of villagers in the retention of ordinary household objects destroyed by the tsunami, piles of rubble left in the midst of the village, and other untouched markers of the tsunami event. Hastrup suggests that "the prominence given by the survivors in Tharangambadi to objects gone, damaged and replaced should perhaps be seen as a function of materiality rather than an indication of this category of loss being the most hurtful. … [S]peaking of the *things* affected by the

tsunami might be just a way of conveying the pervasiveness and reality of the disaster rather than a symptom of a superficial approach to human tragedy" (p. 125).

Equally refreshing is her treatment of the arrival and distribution of aid in the village. Resisting the disaster research dichotomy that often presumes "a too neat distinction between local (hence 'appropriate') and non-local (hence 'inappropriate') humanitarian support" (p. 82), Hastrup demonstrates how most aid was distributed among preexisting structures of power within the traditional caste system of the village. Villagers therefore employed the same strategies they traditionally have, not in a process of *adapting* to an indirect consequence of the tsunami as manifested in an influx of aid, but by *recovering* their share of the material aid through existing channels by exercising their agency. Hastrup rightly notes that "the simple accusation that organisations bulldoze harmonious local communities builds on a potentially equally patronising view of communities as bounded wholes of passive members with no subjective impetus to act within their given frames" (p. 97).

Although a detailed account of Hastrup's numerous and thoughtful theoretical contributions is beyond the parameters of this brief review, an illustrative example can be found in her articulation of a concept she calls a "composite catastrophe" (p. 99). The notion draws a distinction with the construct of a "convergent catastrophe" as put forth by anthropologist Michael Moseley (1999). According to Hastrup's reading of Moseley, a convergent catastrophe suggests that a disaster's impact is larger if it occurs in a temporally proximate sequence with other disasters. To Hastrup, this construct is linked to the notion of a disaster as a one-time occurrence, contrary to the way the villagers of Tharangambadi experienced the tsunami. Hastrup suggests a more fitting construct as embodied in the notion of a "composite catastrophe," which more accurately reflects the villagers' experience of the tsunami as both an ongoing event and as one in a series of events. In other words, "the recovery process criss-crossed clear or temporal chains of causation" (p. 100) for the villagers, both demonstrating the inadequacy of "convergence" as the guiding construct, and demanding the new conceptual category of a "composite catastrophe."

The development of such fresh conceptual constructs as the "composite catastrophe" and other theoretical gems embedded in superbly written ethnography pushes both environmental anthropologies and anthropologies of disaster to consider new platforms of departure and modes of analysis, thereby placing *Weathering the World* in the category of must read for the serious environmental and disaster researcher.

Andrew Tarter
University of Florida

References

Hoffman, Susanna M. 1999. "After Atlas Shrugs: Cultural Change or Persistence after a Disaster." Pp. 302–321 in *The Angry Earth: Disaster in Anthropological Perspective,* ed. A. Oliver-Smith and S.M. Hoffman. London: Routledge.

Moseley, Michael. 1999. "Convergent Catastrophe: Past Patterns and Future Implications of Collateral Natural Disasters in the Andes." Pp. 59–71 in *The Angry Earth: Disaster in Anthropological Perspective,* ed. A. Oliver-Smith and S.M. Hoffman. London: Routledge.

Yin, Robert K. 2009. *Case Study Research: Design and Methods,* 4th ed. London: Sage.

JOHNSTON, Barbara Rose, ed., *Life and Death Matters: Human Rights, Environment and Social Justice,* 2nd ed., 487 pp. Walnut Creek, CA: Left Coast Press, 2011. ISBN 978-1-59874-339-5.

When the first edition of *Life and Death Matters* came out in 1997 (published by AltaMira Press), it became a defining statement for a certain field of research and advocacy; a field that, while hardly new, had not yet seen such a cohesive and comprehensive survey. Editor and contributors alike have worked hard to bring this new edition into the twenty-first century as a living engagement rather than a historical document, with results that manage to be at once depressing and inspiring. The diverse case studies that made up each chapter of the original edition—from China, Mexico, sub-Saharan Africa, Russia, and the United States, among others—have each been reframed into sections marked "End of the Millennium" and "View from the 21st Century." Authors revisit their case studies in detail, reporting changes local and global since the first edition, and considering how the implications of each case are changed or reinforced in hindsight. At least one new author was brought in for these updates, and many of the chapters reflect significant new research undertaken in the interim. Between the larger chapters is a new set of "snapshots," short examinations of socioenvironmental tensions that are new or have been emerging since the 1990s: war in Iraq and the Sudan, impacts of new food and telecommunications technologies, the changing face of climate change. Few academic books see such major revision of content in a new edition, but it seems like a model that more should adopt.

Johnston's new introduction offers some insight into the depth and scope of this type of human-environment research and the way it has adapted to changing times. She also supplies an interesting discussion of the context that framed the initial publication of *Life and Death Matters*: the Rio Earth Summit of 1992, the UN Draft Principles on Human Rights and the Environment, Robert Kaplan's *The Coming Anarchy* (1994), and President Clinton's commitment to limiting arms proliferation, promoting environmental justice at home, and righting (some) wrongs of US interventionism abroad. As presented here, the litany makes UN efforts and the Clinton administration seem quite rosy, which does not quite resonate with the way many of this book's readers felt about those institutions at the time. The Clinton administration's support for unfettered free trade in the form of NAFTA, for example, or the UN's tepid response to genocide in Rwanda, seemed more typical representations. The positive spin on institutions of the 1990s in the introduction seems oddly anachronistic given the sharp criticisms of the same found in most every chapter of the original text.

Any anachronism here, though, says less about the book or its editor than about the times in which we live. Many of Clinton's harshest critics, for example, have come to long for the days of his administration in the decade that followed. Even the more misguided efforts at development in the Third World decried in the first edition lose some of their evil luster when we look back from here in the midst of financial collapse and endless global war on terror. A critical reader might point out that some projects or trends represented as threats (state-managed common property regimes in Southwest China, for example) have not exactly manifested on the scale suggested by the original edition, while we received little to prepare us for arguably much more significant contemporary problems like the global war on terror. In part, this is the nature of case studies, which can never cover all the potential scenarios and outcomes, but it also makes a point about the types of environmental social science. Some disciplines and fields of study are necessarily predictive; global geopolitics and global environmental problems particularly require this kind of investigation. The work that appears in *Life and Death Matters* is no less valuable for not being predictive. The studies are interpretive and analytical, but in ways that take on significance beyond their particular contexts and histories. This leads to another important aspect of *Life or Death Matters* that may account for its staying power since the 1997 publication.

Nothing about *Life and Death Matters* explicitly presents the book as primarily by or for anthropologists, and of course, it would

be silly to make a disciplinary claim on justice, human rights, or the environment. But the first edition created a particular impact among anthropologists; certainly, it was an influence on me and the kinds of graduate study and research I chose to pursue. *Life and Death Matters* remains a foundational reference for a certain population of anthropologists invested in justice and the environment across lines that often divide us sharply; not just theory versus practice, academic versus applied, but scientific versus interpretive or postmodern as well.

A couple of key passages from editor Barbara Rose Johnston might make clear the particular appeal of the book to anthropologists, and of anthropology in the context of these issues. Her introduction begins with a general paragraph on human rights and the challenge of defining and grounding them. From there, though, it moves more boldly:

> For much of human history social groups developed rules and tools for insuring access to critical resources in ways that allowed survival of the group. Some 2.5 million years ago the ancestral *Homo habilis* emerged in Africa. Some 200,000 years ago, *Homo sapiens sapiens* gathered plants, hunted animals, and lived in small mobile communities in the forests and savannas of Africa. By 40,000 years ago our human ancestors had developed ways to live in the heat of the world's deserts, the extreme cold of arctic and mountain terrains, in shady forests, lush and fertile river valleys, and along the coastal shores. And 10,000 years ago humans had fashioned ways to survive and thrive in every major ecosystem on this planet. They developed ways to cultivate and store food to allow for the lean times. They settled in larger numbers in villages, towns, and cities, where their ideas, values, ways of living, and language grew increasingly complicated and diverse. Unlike other creatures whose response to harsh or varied conditions prompted biological change, humans generally relied on their ingenuity to survive. They created innovative ways to live and communicate and thus passed knowledge down to their children. (pp. 9–10)

What an argument! In one paragraph, Johnston pinpoints an essential claim of modern anthropology, tying culture and sociality tightly to evolution and environment. This sets up an argument about diversity, power, and violence that will run through the rest of this book that is otherwise quite diverse in terms of region, method, and scope. This version of adaptation is a strong candidate for an anthropologist's ground rule for understanding the relationship between cultural diversity and healthy environments. The case studies that follow apply Johnston's approach with remarkable consistency across scales and categories of conflict, a robust demonstration of diversity-as-adaptation in the face of what Johnston calls "biodegenerative problems."

Later, in a "snapshot" on nuclear energy written for the new edition, Johnston introduces another element that strikes me as particularly anthropological, one that also informs the larger project tying the book together: "On the question of whether nuclear power is truly 'green' energy, science is easily manipulated. If green is defined as low-carbon and the temporal frame is limited to the operating life of the power plant, we get one answer. If green is defined as healthy for people and their planet now and in the future, we get a very different answer" (pp. 410–411). An epistemology that decenters science as authority and certainty, contesting received definitions and locating normative claims in much larger and longer scales—again, this is an idea of what anthropology should be for that drew many of us to the field.

The other authors in this volume, whether they take up the theme explicitly or not, carry that anthropological thread throughout like a vision statement, even a call to arms. Johnston and the volume's contributors are to be applauded for making such a strong argument with and for anthropology, and for bringing it into the twenty-first century, crises and all, as strong and vibrant as ever.

Adam Henne
Global & Area Studies Program
University of Wyoming

References:

Kaplan, R. 1994. The Coming Anarchy. *Atlantic Monthly* 273(2): p. 58.

KIRBY, Peter Wynn, *Troubled Natures: Waste, Environment, Japan,* 264 pp. Honolulu: University of Hawaii Press, 2010. ISBN 978-0-82483-428-9.

Troubled Natures places its readers squarely in the middle of a new park on the west side of Tokyo. This park is well groomed, tidy, and controlled; it has trees, shrubbery, and flowers planted according to aesthetic standards; it boasts sculpted hillocks, a sports field, a playground, a recreation center, and a rock garden, ringed all by cherry trees. But this park, an oasis of nature within the urban press of Tokyo, rests atop a waste transfer facility where refuse from the megalopolis is trucked in, compacted, and then transferred back out to landfills. The trees and shrubs obscure the truck entrance to the facility, and the hillocks muffle the sounds of the large automobiles lumbering down the streets carrying waste. *Troubled Natures* jumps into the heart of this juxtaposition—of manicured nature sitting atop produced waste—to argue, consonant with a venerable line of anthropology (e.g., Lévi-Strauss 1969; Ortner 1974), that nature is never merely natural. This book contests the idea that the environment is a given, ahistorical, and asocial thing in the world, the stable meaning of which might serve as a universal basis for comparison across times and places. Instead, *Troubled Natures* pushes toward an examination of the social and historical struggles that vest nature with authority, an examination turning specifically on the following questions: Through what embodied interactions are these natures—both the sculpted and groomed nature of the park, and the nature of the waste it rests upon—produced? How does the environment articulate with other elements of social pollution and purity? How, in those wider assemblages, does nature achieve force and authority in people's lives?

Kirby approaches these questions by focusing on two neighborhoods in Tokyo: one involved in a political struggle against toxic effects of the subterranean waste facility, another swept by anxiety-producing media representations of urban pollution. Taking these two crucibles of cultural politics as his base, Kirby moves, across the book, beyond a consideration of waste as mere urban pollution into the "wastescape" of social and ethical notions of purity as well as production, pursuing the meaning of environmental issues within a broader context of urban and individual hygiene, social marginalization and racist self-imaginings, sexual reproduction and economic decline, and the prospects of development and sustainability.

Kirby defines his key term, wastescape, as "a subjective field, shaped more by sociocultural attitudes than by territorial features such as rivers, coastline, mountain ranges, and administrative boundaries" (p. 7) or again as the "commingling between cognitive and concrete landscapes, this anxiety over imagined toxins on the one hand and actual defilement and devastation on the other" (p. 68). With echoes of Appadurai (1996), the term wastescape calls attention to the interaction among different social fields, dispersed yet interconnected, that transverse geopolitical boundaries. Kirby's analysis is attentive to the call for breadth but runs up against longstanding divisions between the subjective and the objective, meaning and reality, or, more troubling for this book in particular, culture and nature.

In mobilizing Appadurai's notion of "scapes," Kirby opens the possibility of recasting such dichotomies in two ways. First, Appadurai takes care to specify the materiality of "imaginary" scapes; they are not floating above or mechanically juxtaposed against a more "real" substrate (1996: 45ff.; see Pedersen 2008: 59ff. for relevant critique). Indeed, the floating world exists in and through its own materiality (see Thompson and Harootunian 1992). Moore, Kosek, and Pandian's (2003) work on race and nature likewise sets such oppositions in dynamic and historical

tension, allowing the analyst to pursue not simply "meaning," and its avatars in etymology (e.g., Kirby: ch. 4; see Martinez 2005 for pertinent critique), but the material and symbolic processes through which the very substance of "nature," "environment," and even "Japan" is manifested, remade, and contested. In such an analysis, one would not have to choose between the poles of the dichotomy (say for example choosing the subjective above the objective, as Kirby urges in his introduction and conclusion) but instead could set them in dialectical tension. Second, Appadurai proposes "scapes" in recognition that "landscapes of group identity … are no longer tightly territorialized, spatially bounded, historically unselfconscious, or culturally homogenous" (1996: 48). Such a logic pushes the analyst to examine how and in what practices "Japan" comes to be relevant, rather than assume it as a given context for cultural patterns, as Kirby at times does (e.g., pp. 9, 68, 85), despite his admonitions (p. 195) and analytic moves (pp. 69ff.) to the contrary. In voicing Appadurai's notion of "scapes," Kirby sets us up to ameliorate his critique of comparative studies reliant on universal definitions of environment (p. 195), open the boundaries of Japan, and allow for a stronger transnational analysis of waste production and management (see Povinelli 2006: 16 on the distinction between comparative and transnational analyses).

Kirby pursues a creative panoply of social issues—a corvine epidemic, entrenched social exclusion, the flagging reproduction of families and the national economy, individual recycling efforts, and the pursuit of "sustainable development"—threaded together as common backdrop to elucidate the meaning of waste. Equally tying this panoply together, however, are the troubles of production—a consistent compulsion to produce, an inability to control the unexpected fecundities of that production, anxieties of production found inadequate. As Japan's economy boomed, consumption increased, aggressive murders of crows came to litter the Tokyo metropolis; dioxins released in factories that power economic growth are found to inhibit human reproduction. The production of nature doubles back on itself. These troubles rely on key metonymic links of bodies to environments (e.g., Tsubo-san's delicate physical sensitivity to the degradation of her surroundings [pp. 28ff]) and the family to the nation. Such themes thread through the entirety of the book, weaving together Kirby's wide-ranging analysis.

Troubled Natures is a vivid ethnographic compendium of some of the ways in which nature and waste become meaningful in Japan. Rendered in graceful and compelling prose, this work charms with vibrant descriptions—of a dead cat taken as symptom of environmental degradations, of the hidden filth of unseen toilet drains, of a golf course built on an island of trash venting flammable fumes—and is studded with erudite considerations of neighborhood life, Japanese exceptionalism, and classical anthropological concerns with purity and pollution. It provides a deeply cultural understanding of the meaning of environment in context. Published alongside the aftermath of the catastrophic environmental disasters of 11 March 2011, *Trouble Natures* is a timely demonstration of how nature, as a human product, is troubled.

Joseph Doyle Hankins
Department of Anthropology
University of California at San Diego

References

Appadurai, Arjun. 1996. *Modernity at Large: Cultural Dimensions of Globalization.* Minneapolis: University of Minnesota Press.

Lévi-Strauss, Claude. 1969. *The Elementary Structures of Kinship.* Ed. R. Needham. Trans. J. H. Bell and J. R. von Strummer. Boston: Beacon Press.

Martinez, Dolores. 2005. "On the 'Nature' of Japanese Culture, or, Is There a Japanese Sense of Nature?" Pp. 185–200 in *A Companion to the Anthropology of Japan,* ed. J. Robertson. Hoboken, NJ: Wiley Blackwell.

Moore, Donald, Jake Kosek, and Anand Pandian. 2003. "The Cultural Politics of Race and Nature: Terrains of Power and Practice." Pp. 1–70 in *Race, Nature, and the Politics of Difference,* ed. D. Moore, J. Kosek, and A. Pandian. Durham, NC: Duke University Press.

Ortner, Sherry. 1974. "Is Female to Male as Nature Is to Culture?" Pp. 68–87 in *Woman, Culture, and Society,* ed. M.Z. Rosaldo and L. Lamphere. Stanford, CA: Stanford University Press.

Pedersen, David. 2008. "Brief Event: The Value of Getting to Value in the Era of 'Globalization.'" *Anthropological Theory* 8, no. 1: 57–77.

Povinelli, Elizabeth. 2006. *The Empire of Love: Toward a Theory of Intimacy, Genealogy, and Carnality.* Durham, NC: Duke University Press.

Thompson, Sarah, and Harry Harootunian. 1992. *Undercurrents in the Floating World: Censorship and Japanese Prints.* Seattle: University of Washington Press.

MCADAM, Jane, ed., *Climate Change and Displacement: Multidisciplinary Perspectives,* 274 pp. Portland, OR: Hart Publishing, 2010. ISBN 9-781-84946-038-5.

The prospect of massive climate change-induced population displacement is a subject of deserved concern as well as clumsy sensationalism, pervasive uncertainty, and conceptual disarray. Legal scholar Jane McAdam brings together scholars from a range of disciplinary backgrounds to offer a more sober and staid, though still compassionate, analysis of this burgeoning threat.

Many of the volume's chapters aim to complicate activist and journalistic assumptions that climate change will straightforwardly create "environmental refugees." Geographer Graeme Hugo warns that statements to that effect, while well-intentioned, ignore the multiple drivers of displacement, homogenize vastly different migration experiences, marginalize in situ adaptation options, and frame climate change-induced migration as a wholly new challenge instead of building on existing migration corridors and mechanisms. Geographers Jon Barnett and Michael Webber stress that involuntary climate change-induced migration may actually increase vulnerability to climate change, while voluntary migration may constitute a diversification of livelihood strategies that allow those left behind to weather the storm. McAdam emphasizes that, notwithstanding the enormous journalistic and scholarly fascination with the prospect of stateless people resulting from vanished island nations, in actuality it is highly unclear that such a legal situation will arise.

Other chapters are more concerned with predicting the obstacles and hardships that climate change migrants will face. Legal scholar Walter Kälin examines the "normative gap" that exists for such migrants: they do not fit easily into any recognized legal category, and thus under the current system will be entitled to, at best, provisional and uncertain assistance. Geographer John Campbell reveals the double cost of climate change displacement in the Pacific Islands: given the social, cultural, and spiritual importance of customary land to Pacific Islanders, displacement in this region will entail great hardship not only for migrants but also for those who are asked to cede their land to the newcomers. Physical and mental health experts Anthony McMichael, Celia McMichael, Helen Berry, and Kathryn Bowen catalog the multiple impacts of climate change on human well-being: direct effects such as deaths from heat waves, as well as indirect effects such as malnutrition caused by crop failure and the psychophysical ravages of forced movement and foreign living conditions. Development studies scholar Roger Zetter argues that conceptual puzzles of causality and voluntariness will create formidable legal obstacles for future environmental migrants. Psychologist Maryanne Loughry, paying special attention to the low-lying Pacific Island nation of Kiribati, argues that the emotional damages of climate change, and the duty of therapeutic assistance, begin not with movement but with the anticipatory fear of it.

Still other chapters are pragmatic in focus. Moral philosopher Peter Penz argues that harm in general, not involuntary movement in particular, should be central to climate change ethics, and therefore states' moral obligation to those hurt by climate change should take the form of a global insurance scheme that supports either migration or adaptation as locals see fit. Political scientist Lorraine Elliott critiques the "security risk" discourse of climate change, in which climate change is presented primarily as a threat to rich countries and government power rather than poor countries and human well-being. Sociologist Stephen Castles's afterword condenses the chapters into nine lessons and policy implications suitable to a disillusioned post-Copenhagen world in which hopes of curtailing climate change have largely vanished.

Inevitably, some readers will come away disappointed. Environmental justice advocates may wish for more writings from a critical perspective: only Penz, Elliott, and Castles focus attention on the moral culpability of industrial nations. Mitigation advocates may take issue with Castles's characterization of greenhouse gas reduction as a hope now dashed and unattainable. Anthropologists and other proponents of bottom-up particularism will be displeased to find not a single detailed case study of a society coping with looming environmental exodus. Even chapters purportedly focused on individual societies (for instance McAdam's and Loughry's chapters on Kiribati and Tuvalu) contain remarkably little in the way of local voices or day-to-day lived realities; ethnographic insight is scant. McAdam in her introduction emphasizes the importance of attending to grassroots voices, and Castles pleas in his afterword for micro-level studies, but the volume's contributors do not truly take up the call.

The volume also somewhat awkwardly straddles specificity and eclecticism. Its focus on the Pacific region feels inconsistent and accidental, and is not effectively incorporated into the book's mission statement. Those who do not specialize in the Pacific may find the frequent reference to that region distracting and repetitive (and perhaps inappropriate, considering the much larger numbers who will be displaced in Asia and Africa), while Pacific Island specialists will find the coverage severely gap-ridden. For instance, as a Marshall Islands scholar I was puzzled to find that country time and again left out of the discussion, even when examining disappearing island nations. Nor was any attention paid to low-lying islanders in the Federated States of Micronesia, Tokelau, or French Polynesia, or to the particular migration possibilities afforded by ties to France or Compacts of Free Association with the United States.

Nevertheless, there is a great deal to admire in this volume. The range of disciplines is commendable. The presentation, with virtually no lapses, is clear and streamlined. Most contributors are leading experts and their authority is everywhere evident. Helpful policy recommendations are offered throughout. Perhaps the volume's greatest asset is its rare ability to acknowledge the gravity of the problem while also undercutting the simplistic and unhelpful alarmism so endemic to this issue. McAdam's volume is a valuable step in coming to grips, both theoretically and practically, with a problem both severe and poorly understood.

Peter Rudiak-Gould
McGill University

MENZIES, Charles R., *Red Flags and Lace Coiffes: Identity and Survival in a Breton Village,* 160 pp. Toronto: University of Toronto Press, 2011. ISBN 9-781-44260-512-1.

Charles Menzies's ethnography *Red Flags and Lace Coiffes: Identity and Survival in a Breton Village* explores how the people of the Bigouden region of Brittany, France, transitioned from being subsistence level agriculturalists in the 1800s to working in sardine canneries and becoming artisanal fishers in the twentieth century. Menzies traces the economic development of the northwestern region of France through providing rich historical and contemporary data. Menzies explores the cultural context within which a local fishery emerged and developed.

The book addresses how social class, gender, and kinship are implicated in the struggles for survival among fishers. In tracing the growth of the fishing industry in the Bigouden, Menzies demonstrates how family-based fishing enterprises continue to survive in the face of challenging odds posed by the forces of globalization. He argues that economic globalization of the late twentieth century coincided with the weakening of the French state's capacity to control local economic conditions and regulate how fishers carried out their activities.

In the book, Menzies traces strikes and protests held by French fishers in the early twentieth century, who wanted to rid the Bigouden of imported fish and to promote local purchasing practices, and the resulting years of social upheaval. Later, the author demonstrates how in the early 1990s fishing incomes in the Bigouden fell 30–40 percent due to European Union policies that allowed fish from nonmember states to enter the marketplace and that cut minimum price supports. In response, many fisher wives began working outside of the home to supplement their husbands' drastically reduced incomes. As working class people joined together in political protest and in caring for their families, a collective identity emerged.

The book's title refers to two important symbols of political struggle: red flags and lace coiffes. Early struggles of international workers are represented by the red flag, which was carried by cannery strikers in the Bigouden in the early twentieth century. The lace coiffe, a tall, white cylindrical hat worn by women that appeared after World War I, began as a symbol of local identity when the region was transitioning from a rural agricultural community to an industrial capitalist economy.

Menzies presents his reader with a varied and complex macropicture of the region, its economic and political struggles, while complementing that with a detailed micropicture of the lives of the fishers at sea and at home. He stresses that fishers are exploited at the levels of the household, the factory, and the fishery. Fishing crews are torn between a notion of egalitarianism in crew relations based on kinship relations versus the structural reality that boat owners and skippers wield considerably more power than do crew members. The author, who worked as a commercial fisherman in Prince Rupert Bay Canada for many years, has both a personal and a professional interest in the lives of his research subjects.

The ethnography is quite short; it is only 129 pages. Yet, the author packs those pages with a lot of useful data. He includes pictures of the fishers, charts on fishing fleets and their characteristics, a glossary of key terms, as well as provides concise vignettes and stories of particular fishing families. Menzies presents several detailed case studies including one of a deckhand and his family, another of a boat owner and his family. The work includes a detailed explanation of the fishing equipment used, types of jobs and fishing vessels in use, and the hours involved in such work. Menzies effectively immerses the reader into the world of artisanal fishers in Southern Brittany.

Yet there are areas of weakness in the text. The book's argument is not clearly laid out at the beginning of the ethnography so that it may get lost to the reader. By the time I reached Part 2 of the book, I remained unclear as to the book's overall argument. This could be improved if the author's unifying concern in writing the ethnography was highlighted more frequently throughout the text.

Generally, the writing is clear and comprehensible, although small sections on economics and the role of the state might be difficult for a layperson to grasp. Because a lot of Marxist economic language is used throughout the text, the content becomes a bit dry and difficult to understand at times.

As a reader, I would have liked to have seen more interview material, especially with the fishermen themselves. Menzies does include a lot of the wives' perspectives, which enriched the ethnography.

Although I found the explanation of the history of the symbolism and importance of both the red flag and the lace coiffe fascinating, I was not convinced of the connection of these potent symbols to the struggles the fishermen faced. Conversely, Menzies provides a fascinating explanation of the transformation of the meaning of the lace coiffe over time as it was worn by various actors in different historical periods and acquired new meanings.

The focus on gender, specifically on the role of women, is relatively brief in the text and could be expanded upon. The author discusses that a working class masculinist identity exists for men, but he does not offer a parallel explanation for working-class women, who admittedly are not the focus of his research.

The book is aimed at introductory anthropology courses and for the most part, this is a reasonable audience for it. The text addresses concerns of the fields of anthropology, the environment, economics, globalization, and history and could be used in a variety of introductory classes. Nevertheless, I would use caution in assigning this book to an introductory anthropology course if the students do not have a strong background in and understanding of economics.

Overall, Charles Menzies has created a rich text that exposes the reader to the hard work and dignity embodied in the labor, political protests, and survival struggles of fishermen and their families in the Bigoudennie from the late nineteenth to the early twenty-first century.

Sharla Blank
Washburn University

MORAN, Emilio F., *Environmental Social Science: Human-Environment Interactions and Sustainability*, 232 pp. Oxford: Wiley-Blackwell, 2010. ISBN 978-1-40510-574-3.

Readers of this journal will be all too familiar with the syllogism that begins with the challenges of the Anthropocene, moves on to the stovepiped insularity of the academic land-scape, and concludes that successful interdis-ciplinary collaboration is both indispensable and terribly hard to achieve. We are all spe-cialists in our own fast-moving fields; how are we to find the time to learn our colleagues' lan-guages well enough to make collaboration pos-sible? This dilemma occurs at all scales, from designing interdisciplinary majors to the for-mation of international steering committees.

Henceforth my answer to this challenge will be Moran's new book, and I intend to keep a stack on hand for both colleagues and stu-dents. The book begins by walking the reader through three topics: a succinct account of cultural evolution from the Paleolithic to the Anthropocene; a lucid overview of the array of environmental challenges now confronting us; and a very helpful history of the creation of the International Geosphere Biosphere Program, the formulation of "Grand Chal-lenges in the Environmental Sciences" by the National Research Council in 2001, and the specific questions and approaches that have moved to the forefront of sustainabil-ity research. Hardly a sentence in the first 24 pages would escape the yellow highlighter of a beginning student or a colleague en route to their first conference on sustainability or global change.

But that is just the overture. Chapter 2 on "theories and concepts from the social sci-ences" is aimed at natural scientists; chapter 3 walks social scientists through "theory and concepts from the natural sciences." Once again the highlighters will be very busy: Moran's explanations are models of clarity. This is the view from 30,000 feet, but Moran tells us enough to capture the potential rel-evance of Boserup, Chayanov, household life cycles, and decision theoretic concepts on the social side, and niche construction, ecosys-tem services, and island biogeography on the natural science side. Forty-eight pages of ref-erences, and an extensive and well-organized index, make it easy to take the next step after reading these chapters.

As I read, I was reminded of my experi-ence of the success of the Santa Fe Institute (SFI) in catalyzing crossdisciplinary collab-orative research. New people are invited to workshops at SFI all the time, and new collab-orations are what keep the institute going. But there is no perceived need to proselytize on the advantages of interdisciplinary research. Instead, each newcomer is faced with the challenge of describing their research in a way that will interest colleagues from several disciplines. Listening to these presentations and conversations, it is common to encoun-ter ideas from other fields that prompt one to wonder if they might shed new light on one's own questions. Reading chapters 2 and 3 of Moran's book, sooner or later almost anyone is bound to find something that will prompt them to think, "Hmmm, can I use that?"

The next three chapters provide an enlight-ening introduction to Moran's major field, multiscale spatial analysis using remote sens-ing and GIS. But Moran is an anthropologist, so his questions soon move beyond pixelated images to the patterns of human-environ-mental interaction that produce them, with his own celebrated work on the deforestation of the Amazon as the centerpiece. The last two chapters introduce agent-based model-ing and models of decision making, and the book concludes with an eloquent perspective on the present and future state of sustainabil-ity science. Here I have a quibble: Moran has little to say about several areas of sustainabil-ity research that arguably deserve a little more ink, such as resilience theory and ecological economics. So along with this wonderful book, I will point interested students and colleagues towards Marten Scheffer's (2009) work on critical transitions, and Dove and Carpenter's magnificent reader on *Environ-mental Anthropology* (2007).

J. Stephen Lansing
School of Anthropology, University of Arizona
Stockholm Resilience Centre
Santa Fe Institute

References

Dove, Michael, and Carol Carpenter. 2007. *Environmental Anthropology: A Historical Reader.* Blackwell Anthologies in Social and Cultural Anthropology. Oxford: Wiley-Blackwell.

Scheffer, Marten. 2009. *Critical Transitions in Nature and Society.* Princeton Studies in Complexity. Princeton: Princeton University Press.

NEWING, Helen, *Conducting Research in Conservation: A Social Science Perspective,* 376 pp. New York: Routledge, 2011. ISBN: 978-0-4154-5791-0.

This textbook is a useful sourcebook for advanced undergraduate and beginning graduate students. It is clearly written, easy to follow, and methodical in its description of basic qualitative and quantitative social science methods for field research with rural populations. Newing, with help from contributors Christine Eagle, Rajindra Puri, and C.W. Watson, systematically describes the nuts and bolts of field research from A to Z in 18 chapters, divided into five sections: Planning a research project; Methods; Fieldwork with local communities; Data processing and analysis; and Writing up, dissemination, and follow-up. Newing covers research design, the specifics of different research techniques (e.g., participant observation, questionnaires, domain analysis and pile sorting for cognitive knowledge, participatory mapping), ethical dilemmas, data analysis (including use of statistical analysis), report writing, and dissemination of research. Each chapter ends with a summary of the major points covered and references for further reading. A strength of the book is the comprehensive treatment of both quantitative and qualitative methods.

The intended audience for the textbook is environmental conservation professionals who are not familiar with social science approaches. Newing wants to make the case that including social science research in conservation work is crucial for more effective policy development and practical application. In the introduction, Newing states that the field of conservation is dominated by biologists, but that given that people are the "drivers" of environmental change, there needs to be a better understanding of the social, economic, and political systems that impact environmental conditions. Indeed, there can be no question that conservationists are seeking and openly welcoming integration of social science perspectives—a considerable sea change from a decade ago. The preeminent conservation biologist Kent Redford (2011), for example, recently wrote a commentary in the journal *Oryx,* in which he calls for increased strategic collaboration with social scientists. Redford, who was once a staunch critic of social scientists who advocated for the inclusion of rural peoples in conservation strategies, seems to have significantly changed his perspective, although he continues to "correct" social scientists' critique of conservation practice.

In this sense, Newing's defense and detailed explanation of social science—particularly anthropology—methodologies can perhaps go a long way to calming conservation professionals' anxieties that social scientists are a fuzzy lot, more interested in feel-good relations with the locals than in true protection of biological diversity. However, I find it troubling that Newing believes that her audience can confidently embrace social science research once they grasp the fundamentals of research methods and techniques, as presented here. Although she specifically states that the textbook is intended to give conservation professionals a basic understanding of social science without the expectation that they will practice it, the book's clinical approach leads one to think that if one can master the techniques, one can apply them. Indeed, university programs for training conservation practitioners increasingly bill themselves as interdisciplinary, guiding students to take a mix of biology, ecology, and social science courses. My experience has been that while students from such programs are more sensitized to the concerns and cultural practices of rural populations—very few programs offer training in urban conservation—they still cannot fully engage in social analysis, a skill gained from deeper disciplinary training that embeds understanding of social behavior in comprehensive familiarity with comparative material, major theories, and associated methodologies. Lacking such capacity, practitioners who want to engage with local populations find themselves entangled in the complexities of human social relationships, particularly the "frictions" (Tsing 2004) that result from unequal power relations at the intersections of local and global processes. They may then get frustrated and once again eschew involvement with local folk in securing conservation goals.

I believe that this book will be more useful for applied anthropology programs than interdisciplinary programs for conservation biologists. Applied anthropologists with a dedication to working on environmental concerns are in short supply. There are now several university anthropology programs with a special focus on environmental or political ecology themes. These would be well served to teach comprehensive methods such as those compiled in Newing's textbook. We urgently need well-trained social scientists who are willing to collaborate with conservation biologists for the common purpose of protecting biological diversity and empowering local peoples to sustain lifeways in their own fashion.

Alaka Wali
Division of Environment, Culture, and
Conservation
The Field Museum of Natural History

References

Redford, Kent H. 2011. "Forum: Misreading the Conservation Landscape." *Oryx* 45, no. 3: 324–330.

Tsing, Anna Lowenhaupt. 2004. *Friction: An Ethnography of Global Connection*. Princeton: Princeton University Press.

PARR, Joy, *Sensing Changes: Technologies, Environments, and the Everyday, 1953–2003*, 304 pp. Vancouver, BC: UBC Press, 2010. ISBN 9-780-77481-724-0.

Sensing Changes by Joy Paar is an innovative study of the hardships and changes experienced by several very different Canadian communities as the result of state sponsored megaprojects or, in the case of Walkerton, Ontario, by a national health disaster that many blamed on provincial government cutbacks. The case studies describe the displacement of a farming community in Gagetown, New Brunswick, by a NATO military base in the early 1950s; the relocation of the community of Iroquois, Ontario, in the late 1950s to facilitate the expansion of the St. Lawrence Seaway; the construction of nuclear reactors in communities in Ontario in the early 1960s and in New Brunswick in 1982; the displacement of farming communities in the Arrow Lakes region of British Columbia in the late 1960s due to the construction of Columbia River Treaty dams; the construction of a heavy water production facility on the shore of Lake Huron in Ontario in the same location as the Douglas Point Nuclear Power Station; and finally, the contamination of the water supply in Walkerton, Ontario, in 2000, that caused seven deaths and nonfatal illnesses for another 2,000 people.

The megaproject status of all but one of these case studies provides an obvious basis for their inclusion in the volume, but the theoretical framework Parr deploys is applicable to a much broader range of phenomenon. Her study is informed, in part, by the field of critical environmental history and from that perspective Paar raises troubling questions about the history of industrialization in Canada and the arbitrary manner in which national and provincial governments have routinely, and without consultation, demanded the ultimate sacrifice from some communities "for the good of all." In the Arrow Lakes region of British Columbia, for instance, local residents were not consulted before Canada and the United States mutually agreed to permanently flood their lands in order to create a storage reservoir to augment the downstream capacity of American hydropower facilities and, ironically, to improve flood control measures in the lower portions of the Columbia River Basin in the US.

However, while a critical environmental voice is present in this volume, other theoretical interests are more prominent. Of special importance are theories of embodiment as developed within the disciplines of anthropology, sociology, history, and philosophy. Paar's epistemological position on embodiment is consistent with the work of Pierre Bourdieu but opposed to the work of Foucault and Butler among others, who, she argues, treat the body as metaphor rather than a source of knowledge in its own right (pp. 8, 20). She is also critical of social constructivism which she represents as based on the premise that "meaning precedes experience" (p. 12)—that knowledge construction occurs through social interaction and is then written on the body, as opposed to the idea that the body itself is a site for knowledge construction, independent of social process. Although her argument might stand as a useful corrective for those who believe, rather too literally, in Derrida's claim that "there is nothing outside the text" (Derrida 1976:163), Paar herself assigns a truth value to bodily experience that may be overstated, as when she writes in her introduction that "the body is a synthesizing instrument that defies the categorical and linear discipline of language and science" (p. 4). Given the amount of attention she devotes to these arguments in her introduction, some readers may be surprised and disappointed by the limited ability of her case studies to speak to these distinctions. Her study of the Gagetown military base, for instance, is based entirely on archival research, not personal observations. She is able to excavate some information from archival sources about the embodied nature of rural, farming livelihoods in that setting. And in keeping with her intention to not overly privilege the visual above all

other senses, she provides descriptions of the smells and sounds of rural lives, the physical rhythms of daily work, and an orientation to place and landscape rooted in several generations of occupation by the Euro-Canadian families being described. Paar has done extensive firsthand research in most of the communities she describes, but her focus in those case studies as well is mainly on what she refers to as "sensuous knowledge"— descriptive details that enrich the story and deepen the reader's appreciation for what was lost when families were forced to move. These descriptive details do not, however, touch the deeper, often unconscious, unspoken and unspeakable forms of bodily knowledge that she refers to throughout her introduction (see especially pp. 8–11).

Paar's descriptions of embodied knowledge do link extremely well to a further important theme of the book, the relation of the senses to technological change. As she states in her introduction: "As human interactions with environments, technologies, and the everyday have changed, the senses have been tuned, over time, to bring different qualities to human bodies" (p. 11). A recurrent theme in her case studies is the replacement of habituated sensory knowledge with new ways of knowing that require not only the "retuning" of the senses to changed ecologies, but a greater dependence on technological ways of knowing. Workers in the first nuclear power plants in Canada, for instance, had to replace the auditory and visual surveillance practices learned in other work sites, with "instrumental surveillance" in order to assess the risk level of soundless, invisible radiation (p. 62). People living in Inverhuron, Ontario, near the heavy water plant described in chapter 6, discovered that, while smell could alert them to the presence of small amounts of the hydrogen sulfide periodically emitted by the plant, a large and potentially fatal level of emissions entirely extinguished their sense of smell. In Walkerton, where seven people died because of exposure to the E. coli that had contaminated their water supply, the media alternately blamed the incompetence of local water managers and the negligence of a provincial government that had downsized the agencies responsible for drinking water quality. But as Parr's study demonstrates, many citizens of Walkerton, including its water managers, were wary of chlorination because of their appreciation for "clean" water. As a result, it was not applied at the levels necessary to protect them. In the aftermath of the tragedy, scientific measurement and the presence, rather than absence, of chlorine, came to define "clean" water for many Walkerton residents.

This artful and eloquently written collection of case studies may not provide compelling evidence for the corporeal embodiment theory outlined in the introduction, but it does link embodiment theory to technology studies and critical environmental history in an innovative and thought-provoking manner. It is a particularly strong contribution within the field of Canadian environmental history and its theoretical orientation, methodological innovation, and incorporation of "sensuous knowledge" merit a wide readership.

John Wagner
UBC Okanagan

References

Derrida, Jacques. *Of Grammatology.* Translated by Gayatri Chakravorty Spivak. Baltimore: John Hopkins University Press, 1976.

RADEMACHER, Anne M., *Reigning the River: Urban Ecologies and Political Transformation in Kathmandu,* 245 pp. Durham and London: Duke University Press, 2011. ISBN 9-780-82235-080-4.

Over the years, I often have walked alongside the rivers of the Kathmandu Valley, carefully treading the garbage-strewn bank of a river, trying to stay upwind of its odor, and disturbed by the toxic, discolored river water that flowed past me and by the lack of an effective government response to its degradation. I mainly go to the rivers because of the temples that line them and affirm their sacred character. The juxtaposition of spiritual purity and secular pollution, embodied in Nepal's Bagmati River, has never ceased to fascinate me, and the cultural significance of the rivers among valley inhabitants is one of their most compelling traits. And so it was with great interest that I picked up Anne Rademacher's engaging new book, *Reigning the River: Urban Ecologies and Political Transformation in Kathmandu.* I found it to be at once a richly grounded ethnography of life along the rivers of the Kathmandu Valley, an insightful development of urban ecology theory, and a compelling portrayal of local environmental advocacy amid political and societal change. It also is very well written. It makes clear that my earlier thoughts about a lack of "government response" to river deterioration missed the point—healthy rivers in the valley will lie more with the engagement of residents and river-focused activists than with office-based politicians.

The book opens with the author's discussion of her presence in Kathmandu and in the study, itself. It situates her life in the city among the various actors of river restoration in Nepal—residents, government bureaucrats, activists, scientists, and squatters. This is helpful and it signals the primary thrust of the book—an explication not just of the biophysical or geochemical degradation of the rivers, but rather of how their decline (and hopeful restoration) lies not in the evolving political milieu of the Kathmandu Valley but within society and the everyday life of people she has known and with some of whom she has lived. It is this personal commentary, intermixing experience, theory, and factual information, that give the book its compelling sense of immediacy and authenticity.

There exist numerous surveys and studies of Nepal's rivers—for wide-ranging purposes, including the study of glacial loss, the analysis of environmental impacts of industry and agriculture, the assessment of energy from dams, and to track changes in cultural identity. The Kathmandu Valley's rivers have been singled out especially for physical study because of their high pollution and sediment loads, and because so many people depend on them for daily life: drinking water, farming, laundry, industry, and even recreation and worship. The author refers in her book to many of these studies, especially in the chapters that recount the hydrology and chemical contamination of the rivers, but always with an eye toward grounding the science studies in the ordinary affairs of human life. In other words, she embeds the rivers' ecologies in the social and household life of the city's residents. This is a major strength of the book. Of particular concern to the author are the new residents who have migrated to the valley and now live in squatter settlements or spontaneous housing colonies along the river banks (*sukambasi*). These riverbank populations play a contentious and poorly understood role in both the rivers' decline and restoration. They are visible reminders of the plight of the city's poor, and an often maligned symbol of its environmental woes. Of particular note for purposes of the book is the often-contentious relationship between the squatters, housing advocates, and Nepal's government—a relationship that first must heal even before the rivers can be made healthy.

A significant portion of the book is devoted to the changing political climate of Nepal and the Kathmandu Valley, with recorded events ranging from the massacre of the royal family in 2001 to the everyday casual disinterest of some government bureaucrats. I could not always follow the linkages the author makes between the recounted political events and changes in the river, beyond a background framing, especially insofar as actual policy is concerned, but the author does make clear how the evolving political climate came to bear on ways in which the state of the valley's rivers were portrayed in the media, on the administrative failures that led to a lack of resolve or effective land use planning, and on the hopes of a nascent democracy for a restored country, city, and river.

In its six chapters (plus a foreword and conclusion), the book is a palimpsest of social and environmental settings, actors, and events acting together in the Kathmandu Valley. Tying these components together, and framing much of the book, is the intersection of the physical and social contexts of river degradation and restoration. In exploring how river pollution in the Kathmandu Valley is, at least in part, a product of metropolitan society and a lack of political will, the book carries the concerns of the Himalayan literature on land degradation and society squarely into the urban realm. Looking forward, most of Nepal's population (and the world's) will be city dwelling, and so this book's efforts to "urbanize" the long-standing concerns of political ecologists and conservationists are timely and welcome. Beyond its theoretical contributions, though, the book stands out as an essential primer for those involved in governance and development in Nepal (and elsewhere).

In the end, the book's main strengths lie in the insightful ways the author grasps the conflicting agendas of persons whose own lives intertwine with the life of the rivers—local inhabitants, environmental activists, development bureaucrats, squatters, housing authorities, media personnel, and politicians, and how the future of the rivers rests on the shared ownership of the urban environment. Therein lay the promises of "urban ecology." The term itself appears in the subtitle of the book. It first shows up in the book in a novel way—the author recounts a conversation with the director of a civic organization in which she was describing her study. He asked her what she meant by "urban ecology." Her best answer, I suspect, is this book.

David Zurick
Eastern Kentucky University

RUTHERFORD, Stephanie, *Governing the Wild: Ecotours of Power,* 250 pp. Minneapolis: University for Minnesota Press, 2011. ISBN 9-870-81667-447-3.

What do the American Museum of Natural History's Hall of Biodiversity, Disney's Animal Kingdom theme park, an ecotour of Yellowstone and Grand Teton National Parks, and Al Gore's *An Inconvenient Truth* documentary have in common? According to Stephanie Rutherford in *Governing the Wild*, they are all forms of green governmentality, particular constellations of knowledge-power seeking to both control nonhuman natures and instruct human subjects in how to properly relate to these natures. As Rutherford summarizes, these different projects are each concerned with "defining and fixing the appropriate relationships between the human and nonhuman through the lens of science, the practice of biopolitics, the making of subjectivity, and the dominion of the commodity form" (p. 203).

In advancing this thesis, Rutherford builds on a growing body of work applying Foucault's popular governmentality analytic—the assertion, in a nutshell, that governance is not exercised predominantly by states but operates throughout the social landscape, addressing all subjects as both point of application and vehicles of transmission—to the study of environmental governance. Rutherford's main contribution to this literature lies in her differentiation among four distinct strains of green governmentality that "Foucault did not anticipate" (p. xx), each exemplified by one of her four case studies. The Hall of Biodiversity's taxonomic exhibits, in this model, represent green governmentality's "scientific dimensions," acting as "a scientized assessment of global nature under threat" (p. xx). The hyperreal experience provided by Disney's Animal Kingdom, by contrast, represents a "corporate" governmentality in its depiction of "nature as simultaneously fantastic and spectacular while donning the mantle of legitimacy by … reimagining itself into a kind of nongovernmental organization" (p.

xxi). Rutherford's Yellowstone/Grand Teton ecotour exemplifies an "aesthetic" governmentality, operating as "an intellectual factory producing the visual grammars of 'pristine' national natures" (p. xxi). Finally, the narrative of crisis and redemption through technological fix offered in *An Inconvenient Truth* represents a "moral" governmentality, promoting "a moral green subjectivity through which to encounter the natural world" (p. xxii).

All these forms of green governmentality are problematic, Rutherford contends, because they highlight certain aspects of the nonhuman natures they represent while obfuscating other, less benign dimensions, thereby presenting "sanitized" depictions and concealing more problematic questions of power and structural inequality, for instance, in the destruction and salvation of "the environment." Time and again, she calls attention to dynamics that conflict with the ostensibly altruistic agenda promoted in the different projects she analyzes, from the underpaid service workers supporting the Yellowstone ecotour to the animals inadvertently killed in the Animal Kingdom's development. In this green governmentality vision, nonhumans are rendered passive and docile (in planners' perceptions if not in actuality) while humans are enjoined to display their singular agency in service of the environment by becoming "the perfect moral green citizen: one that doesn't consume less, but consumes appropriately" (p. xxii)—primarily by purchasing the environmentally friendly products and services on offer in the various venues. In this way, Rutherford claims, her green governmentality projects demonstrate "the amazing capacity of capitalism to innovate and colonize, to expand in this case into nature, which is perhaps something of a biopolitical frontier" (p. 88).

As evidenced in this passage, throughout Rutherford's narrative runs a peculiar tension between governmentality and commodification, biopower and capital. At times she describes her projects as foremost about knowing and controlling nonhuman natures, at others about selling products under the pretense that their purchase contributes to conservation. This, of course, reflects in part a long-standing debate between Marxist and poststructuralist positions generally concerning the chief motives for human action, with Marxists characteristically positing that ideological projects conceal base material interests while poststructuralists hold that discourse—Foucault's "will to know"—can be a motivating force in its own right. In Rutherford's analysis, however, the tension between these positions is left rather unresolved.

The analytical lens that could unite them is neoliberalism, and while Rutherford mentions the term on several occasions her discussion remains little developed—curious given that her references included Foucault's (2008) recently published series of Collège de France lectures from 1979 titled *The Birth of Biopolitics*, in which neoliberalism is described as a specific form of governmentality, contrasted with a "disciplinary" art of governance emphasizing diffusion of ethical norms more commonly associated with the concept. In this description, neoliberalism is not merely a form of capitalism but a particular approach to governing human action in general, primarily aimed at the construction and manipulation of the external incentive structures in terms of which actors make decisions. Neoliberal governmentality has been identified in the type of environmental projects Rutherford describes in their aim to incentivize conservation by conferring, via commodification, an appropriate market value to in situ natural resources such that preservation will be favored over depletion (e.g., Fletcher 2010; Oels 2005), while the same programs have also been described from a more orthodox Marxist point of view as vehicles for the proliferation of profit-seeking neoliberal capitalism (e.g., Brockington and Duffy 2010; Brockington et al. 2008). Rutherford's tacking between explanations in terms of governmentality and commodification likely reflects this dual character of the neoliberal program as exemplified in the various projects she analyzes, all of which display a characteristic

neoliberal preoccupation with market engagement, a process Rutherford points to but neglects to interrogate in depth.

In his *Biopolitics* lectures, Foucault distinguishes two additional forms of governmentality: a sovereign form bent on top-down regulation; and what he calls "art of government according to truth"—ostensive truth concerning the innate order of things as revealed through religious text, for example. It would be interesting to rework Rutherford's analysis in terms of this typology, describing her multiple governmentalities not according to the science-corporate-aesthetic-moral distinctions she develops (and about which I am somewhat skeptical given that all four of these elements are present to different degrees in each of the projects she analyzes) but rather as disciplinary, neoliberal, and truth-based strategies for environmental behavior modification. In this frame, Rutherford's aesthetic and moral forms would be elements of a disciplinary governmentality, her corporate form neoliberal governmentality, and her scientific governmentality a variant of Foucault's truth-based art of government. This might allow for a more nuanced analysis of how these various approaches mix and match in particular projects.

Such a reading would have intriguing implications in relation to Rutherford's conclusions as well. Again following her intellectual mentor Foucault, Rutherford refrains from offering specific solutions to the dilemmas she highlights. Rather, drawing on actor-network theory, she points to the possibility of developing an alternative "relational ethics" that understands interactions between humans and nonhumans not as "watcher and watched," as in the green governmentality register, but as an "encounter" between different "actants" all of whom exercise their own forms of (not necessarily equivalent) agency. This, Rutherford suggests, "would require more than simply looking upon; it would necessitate engagement, response, interaction," taking "into account the economic, political historical, technological, social, biophysical, and cultural practices" obscured within the green governmentality lens (p. 201). Such encounters would then become "an opportunity to know something about the animals rather than our reactions to and visions of them" (p. 202).

This prescription displays a curious slippage between notions of power and freedom, reality and illusion, somewhat at odds with the Foucaultian framework on which Rutherford relies. She claims not to "suggest that there wouldn't be power relations laden" (p. 201) in her hypothesized human-honhuman "encounter." However, there remains a certain sense in which Rutherford seems to posit the possibility of occupying a "space outside of green governmentality" entirely (p. 42). This would seem to run counter to Foucault's oft-repeated assertion that there is no outside of power as he defines it. (It also renders Rutherford's identification of her various projects as exercises in governmentality somewhat superfluous, since from a Foucaultian perspective how could they be otherwise?) Similarly, Rutherford appears to point toward the potential for an encounter that acknowledges the "actuality" (a term she employs on several occasions) of nonhuman natures, belying the poststructuralist assertion (à la Baudrillard, whom she references frequently) that it is precisely in the constructed opposition between real and false that the "fiction" of the real is in fact redeemed. (In this frame, Disney's Animal Kingdom would serve less to obscure an actual nature, as in Rutherford's analysis, than to rejuvenate—through ostentatious hyper-reality—"the fiction of the real in the opposite camp," as Baudrillard describes the function of Disneyland.) Consequently, Rutherford's prescription of relational ethics as an antidote to green governmentality seems somewhat out of step with the rest of her Foucault-inspired deconstruction.

Foucault himself suggests a different response to these issues. At the end of the *Biopolitics* lectures, he speculates concerning the nature of a distinctive "socialist" art of government standing in contrast to all of the other forms he outlines. What if, rather that proposing a transcendence of green governmentality entirely, Rutherford had explored what such an (noncommodifying? antibiopolitical?) art of government might look like with respect to human-nonhuman encounters? Envisioning such an alternative stands as one of the most pressing next steps for an environmental politics seeking to shrug the neoliberal straightjacket increasingly imposed on it by the type of biopolitical projects Rutherford analyzes. Although it does not necessarily provide a roadmap for such a project, *Governing the Wild* does usefully highlight the daunting obstacles that must be overcome for this to be possible.

Robert Fletcher
Department of Environment, Peace, and Security
University for Peace

References

Brockington, Dan, and Rosaleen Duffy, eds. 2010. *Antipode* 42, no. 3. Special issue on *Capitalism and Conservation*.

Brockington, Dan, Rosaleen Duffy, and Jim Igoe. 2008. *Nature Unbound: Conservation, Capitalism and the Future of Protected Areas*. London: Earthscan.

Fletcher, Robert, 2010. "Neoliberal Environmentality: Towards a Poststructuralist Political Ecology of the Conservation Debate." *Conservation and Society* 8, no. 3: 171–181.

Foucault, Michel. 2008. *The Birth of Biopolitics*. New York: Palgrave Macmillan.

Oels, Angela. 2005. "Rendering Climate Change Governable: From Biopower to Advanced Liberal Government?" *Journal of Environmental Policy & Planning* 7: 185–207.

WALKER, Peter A. and Patrick T. HURLEY, *Planning Paradise: Politics and Visioning of Land Use in Oregon*, 312 pp. Tucson: University of Arizona Press, 2011. ISBN 978-0-8165-2883-7.

Planning Paradise is the latest addition to the University of Arizona Press Society, Environment, and Place series, which presents human-environment scholarship with an interdisciplinary appeal. Geographers Peter A. Walker and Patrick T. Hurley's book offers an examination of Oregon's land-use planning system written for an interdisciplinary academic audience as well as practitioners and citizens. Part celebration and part warning, *Planning Paradise* is an excellent example of how social science research can offer lessons for environmental policy.

Walker and Hurley uncover the cultural and political aspects of land use, challenging "notions that planning is merely rational, technical, and objective" (p. 21). In the introduction, they cite Raymond Williams's *The Country and the City* (1973) and Roderick Nash's *Wilderness and the American Mind* (1982) as examples of how "cultural *visions* or ideals" (p. 15) about place "become justifications for policies" (p. 16). Though the authors do not explicitly develop a theoretical framework for their text, those familiar with their collaborative work (Hurley and Walker 2004; Walker and Hurley 2004) and Walker's critical essays for *Progress in Human Geography* (2005, 2006, 2007) will recognize undercurrents of political ecology and cultural landscape studies.

Following the introduction, the authors devote three chapters to an extensive history and discussion of the Oregon Land Conservation and Development Act (SB 100) and several statewide ballot measures (pp. 7, 10, 11, 37, and 49). Chapter 2 reviews Oregon's state-centered planning approach with descriptions of Statewide Planning Goals, Urban Growth Boundaries (UGB), and administrative structures associated with SB 100. Without a flow chart or glossary of acronyms, this chapter is a bit difficult to digest. However, the next two chapters provide greater clarity by animating these policies and structures.

In chapter 3, Walker and Hurley dispel the myth that SB 100 was the product of "environmentalists or left-leaning city folk" (p. 42). They describe the main architect of the bill, Hector Macpherson Jr., as a "cerebral dairy farmer" (p. 46) who was inspired by landscape architect Ian McHarg's *Design with Nature* (1969) and the report *The Quiet Revolution in Land Use Control* (Bosselman and Callies 1971). The authors also uncover a coalescence of political forces that helped pass the bill, including public outreach by governor Tom McCall; nonpartisanship in the state legislature; farmers organizations and labor unions; and anti-California sentiment among voters. In chapter 4, Walker and Hurley show how state ballot initiatives in the early 2000s, which challenged SB 100, represented a shift in public sentiment away from state-centered planning toward individual economic and private property interests. These historical chapters are successful because they offer a wide political-economic lens but remain sharply focused on SB 100 and subsequent ballot measures. In chapter 5, Walker and Hurley argue that while the planning community has dismissed recent challenges as "mere clever political tactics by opponents," measures 7 and 37 expose "serious problems" (p. 113) to the longevity and stability of SB 100. They argue that the planning system has had trouble adapting to changing economic and political contexts and needs to revive its "commitment to meaningful public engagement that characterized the program in its early years" (p. 140).

To illustrate these challenges, the next three chapters present regional case studies. Packed with maps, photos, and empirical detail, these chapters provide place-specific examples that make spatial sense of the state's land-use politics. Since UGBs are a centerpiece of Oregon's planning system, the authors start with a study of the Portland Metro Area, illustrating how rigid definitions of farming in SB 100 create problems for protecting farms that serve urban markets. Chapter 7 focuses on the Metolius Basin, where conflicts over destination resorts in rural places expose a struggle over "*which* public, at *what scale*, exactly, the state planning system serves" (p. 201). The third and final case study shows how resistance in southern Oregon exposes similar scaled conflicts over "*how much* noncompliance" with state initiatives is acceptable (p. 229). Together these case studies illustrate one of the most difficult challenges Oregon's planners face: to foster regional coordination and engage local innovations while controlling interest group politics.

In their conclusion, Walker and Hurley use historical and modern political issues from previous chapters to discuss "whether Oregon's planning system can and will change *enough* to survive in an era that is ... fundamentally different than the time of its birth" (p. 232). They argue that land-use planning needs to move beyond "the 'bright line' between country and city" (p. 254) because it obscures the political interdependency of these places. They also suggest that planners should draw on the historical success of participatory politics, and they point to social science research that suggests public outreach "*makes* informed, engaged, and thoughtful citizens who are far more likely to support sound planning" (p. 246).

This conclusion by way of policy recommendation fulfills the authors' intentions to speak beyond the academy. At an author meets critic session at the 2011 Association of American Geographers conference, the book was criticized for its lack of theory. Though this shortcoming may hinder its use in some upper-level graduate seminars, its focus on political practice makes it an excellent choice for graduate seminars in planning, public administration, or environmental law. Ultimately, the authors may not be concerned if we do not use this book in our classrooms. To court a wider audience they consciously sacrifice theoretical reflection for an emphasis on policy, writing the book for "those who wish to protect Oregon's truly remarkable landscapes" (p. xii). To realize this goal, Walker presented the ideas of *Planning Paradise* to members of the nonprofit 1,000 Friends of Oregon, urging these citizens to think about how they "can make the land use planning system not just less politically vulnerable, but more politically robust" (Beebe 2011). This

message will also resonate with academics who are interested in the practical implications of social science and anyone looking for the policy in political ecology.

Brian Grabbatin
Department of Geography
University of Kentucky

References

Beebe, Craig. 2011. "Rediscovering 'Paradise': Peter Walker on Oregon's Land Use Challenge." *Oregon Stories,* July. http://www.friends.org/about/profiles/Peter-Walker-QA. Last accessed on September 15, 2012.

Bosselman, Fred P., and David L. Callies. 1971. *The Quiet Revolution in Land Use Control.* Washington, DC: Council on Environmental Quality. http://www.eric.ed.gov/PDFS/ED067272.pdf. Last accessed on November 10, 2012.

Hurley, Patrick T., and Peter A. Walker. 2004. "Whose Vision? Conspiracy Theory and Land-Use Planning in Nevada County, California." *Environment and Planning A* 36(9): 1529–1547.

McHarg, Ian L. 1969. *Design with Nature.* Garden City, NY: American Museum of Natural History/Natural History Press.

Nash, Roderick. 1982. *Wilderness and the American mind.* 3rd ed. New Haven: Yale University Press.

Walker, Peter A. 2005. "Political Ecology: Where Is the Ecology?" *Progress in Human Geography* 29, no. 1: 73–82.

Walker, Peter A. 2006. "Political Ecology: Where Is the Policy?" *Progress in Human Geography* 30, no. 3: 382–395.

Walker, Peter A. 2007. "Political Ecology: Where Is the Politics?" *Progress in Human Geography* 31, no. 3: 363–369.

Walker, Peter A., and Patrick T. Hurley. 2004. "Collaboration Derailed: The Politics of 'Community-Based' Resource Management in Nevada County." *Society & Natural Resources* 17, no. 8: 735–751.

Williams, Raymond. 1973. *The Country and the City.* New York: Oxford University Press.

www.ingramcontent.com/pod-product-compliance
Lightning Source LLC
Chambersburg PA
CBHW081108220326
41598CB00038B/7276